2016 공간의 조형미와 최신 건축자재의 트랜드

공간구성과 트랜드

Ⅰ

공간의 조형미와 최신 건축자재의 트랜드

2016 공간의 조형미와 최신 건축자재의 트랜드

공간구성과 트랜드 I

2016 공간구성과 트랜드

공간의 조형미와 최신 건축자재의 트랜드

2016년 공간의 조형미와 최신 건축자재의 트랜드-
2016 공간구성과 트랜드 도서의 모든 저작권과 소유권은
도서출판 아르스에게 있으며, 이 책에 실린 모든 페이지의
내용과 사진이미지는 허가없이 복사할 수 없으며,
위반시 모든 법적 민, 형사상의 책임을 질 수 있음을 알려드립니다.

2016 공간의 조형미와 최신 건축자재의 트랜드

공간구성과 트랜드

Architecture라는 단어에는 다양한 단상과 철학적 사상, 포괄적 의미들이 숨겨져 있습니다. 현대문명의 환경 속에서 건축학적 의미와 실체들, 다양하고 복잡한 현상들과 탄생과 재탄생의 헤아릴 수 없는 경험들, 건축행위 주체들의 각고한 노력과 개념을 포함한 수 없이 많은 건축적 아름다움과 편리함을 경험하고 있습니다. 또 이 건축이라는 단어에는 또 다른 작은 요소들이 포함되어 있습니다. 방, 정원, 계단, 지붕, 창, 등의 내적요소와 그 이상의 외적요소입니다. 산, 호수, 숲, 잔디, 나무, 동물, 돌, 물 등을 포함합니다. 그리고 가족과 직원, 손님, 애완동물 등도 포함됩니다. 또한 예술로 쓰이기도 하는데 공예, 조각, 사진 그림 등이 포함됩니다. 실로 건축이라는 단어에는 너무나 다양하고 큰 의미가 포함됩니다. 건축이 대변하는 실체는 우리의 감각을 통하여 경험되어집니다.

우리는 건축을 어떻게 받아들이고 이해하고 있을까?
건축은 우리가 보는 것의 경험, 느끼고 듣고 냄새 맡는 것을 포함하여, 볼륨의 기하학적 결과 그리고 표면을 이루는 다양한 건축 재료들, 건물의 장식, 색채, 텍스쳐, 연계, 투명성 등의 것, 그러므로 건축물의 모든 것을 일컫는다고 볼 수 있습니다. 두번째는 우리는 무엇인가? 라는 것입니다. 우리가 행위하고 경험하고 무엇인가를 지속하는 것은 건축입니다. 누군가 무엇을 하는 것을 보는 것이기도 합니다. 건축은 이미 언급한대로 살아가는 가운데 이루어지는 거의 모든 과정에 연관성과 관련되어 있습니다. 건축은 어떠한 공간에서 또는 어떠한 장소에서 이루어집니다. 다른공간과 관련된 공간에서 그리고 그밖의 외부공간에서 인접, 대응의 관계에서 이루어지고 그 공간은 기하학적이 아닌 위상학적(Topological)관계를 가집니다. 그리고 건축물을 통하여 움직이거나 들어설때와 같은 지속적인 경험을 만들어 냅니다.

아무리 멋지고 훌륭한 집을 사고 값비싼 가구들을 골라 들여 놓는다고 하지만 그것으로 삶이 아름답거나 자신의 거처가 훌륭해 질수는 없습니다. 우리가 가질 수 있고 원하는 것 모두가 적정한 시간과 역사를 필요로 합니다. 그리고 그것은 우리 스스로 생산하고 가꾸어 낼 수 있어야 합니다. 건축은 그러한 모든 것 가운데 가장 중요하고 다른 어떤 것보다 더 근본적인 무대를 창조해내는 것입니다. 그것은 개인으로써 혹은 한 사회가 건강한 구성원의 내적 안정을 유지하는데 절실히 필요로 하는 것이기도 합니다.

현대를 살아가는 사람들에게 공간으로써의 건축의 개념과 그 건축을 설계하고 건설하는 전문집단에게 있어 눈 앞에 펼쳐신 시각적 풍경들, 눈을 위한 환경들을 디자인하는 것은 모두에게 중요합니다. 우리가 가끔씩 찾아가는 바닷가, 모래사장 그리고 멀리보이는 항구의 불빛까지도 모두 건축을 통하여 구현된 시각적 산물이고 우리가 기대하는 삶의 무대이며, 건축은 우리가 이 세상을 살아가는 동안 지속될 행위입니다. 그러므로 건축은 우리가 언제나 새롭게 창조하고 새로운 다양한 재료들과 조화를 경험하며, 거의 모든 삶의 이벤트와 활동 그리고 그 결과물은 건축이라는 무대에서 구체화 합니다. 건축은 우리가 생산하는 모든 것을 담으며, 문화를 창조하고 유지하는 것이기도 합니다.

따라서, 이 책에 실린 모든 건축물의 건축재료, 건축기법, 공간의 조형적 완성도 등 새롭게 창조된 건축물의 이해는 향후 새롭게 탄생할 건축의 완성도에 조금이나마 도움이 될 것이라 확신합니다.

박제하 / 2016.4.

• 이 책자에 실린 건축물의 건축자재는 보편적으로 용도상 중복적으로 쓰여지는 건축물의 특성상 다음과 같이 픽토그램으로 구분되어 표시되었습니다.

2016 공간구성과 트랜드 I
공간의 조형미와 최신 건축자재의 트랜드

- **C** Material Type - **Concrete**
- **B** Material Type - **Brick**
- **G** Material Type - **Glass**
- **M** Material Type - **Steel & Metal**
- **W** Material Type - **Wood**
- **S** Material Type - **Stone**
- **P** Material Type - **Plastic**

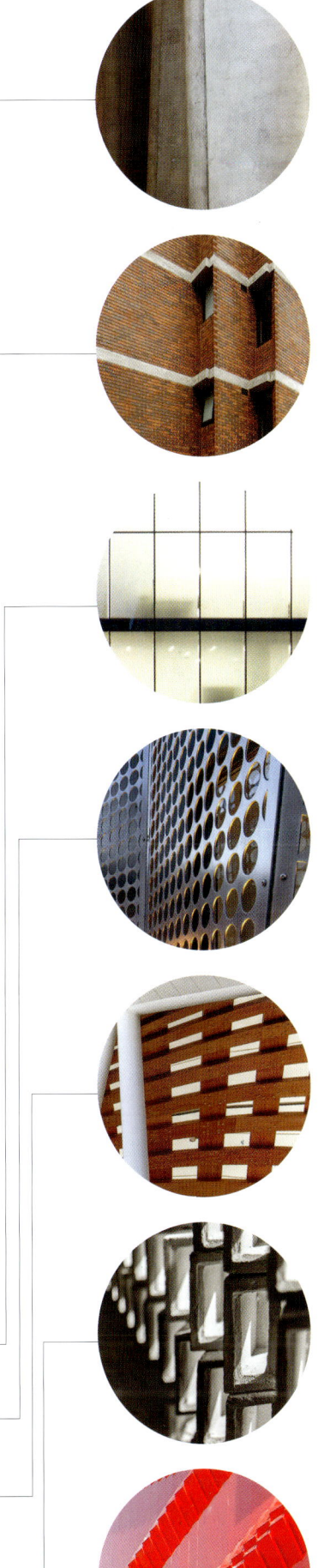

Spring Art Museum

Architects: Praxis d'Architecture
Location: Daxing Qu, Beijing Shi, China
Area: 4700.0 sqm   **Project Year:** 2015
Photographs: Xia Zhi, Jin Fengzhe, Courtesy of Praxis d'Architecture, Zhou Ruogu
Project Architect: Di Shaohua
Design Team: Zhang Xiaodong, Liu Xing, Di Xiangjie, Feng Jiancheng, Feng Shuxian, He Feng, Wang Boyu

Our client for this project is a local who believes in contemporary art, and intended the museum to be a platform to promote young artists. The two main programs of the museum are for art exhibition and artist residence. In the design we sought a piece of architecture that is open, culturally rooted, and arousing a feeling of sublimity – a noble condition of human instinct, which we believe, although through different means, shares the same purpose as the art it accommodates.

Located next to a lake, the site marks the east end of a cluster consisting of art exhibition spaces and artist studios. The lake is to the north, and in the south right adjacent is Song Zhuang Museum. The site is below the road level at south and east and surrounded by parapet at maximum 2m high.

The building massing was resolved into a U shape that recalls traditional triple house courtyard familiar to the local people. The courtyard faces a road on the east side and provides entries to the main exhibition spaces. There is an uninterrupted path from the road level to the courtyard and then all the way to the roof, which facilitates outdoor art display, draws activities in and blurs the boundary between roof and facade.

Raised by 1.6m above the road level in the south and east, the courtyard is roof to the ground floor to be entered from north and west. Two sunken gardens punctuate one level down to reach the earth, allowing big trees to grow out, as well as providing natural light and ventilation to the ground floor that holds some dark spaces, a lounge, and several studios for resident artists. The roof comprises series of terraces at different levels, and the height difference allows skylight into major exhibition spaces in which the ceiling profile corresponds to that of the roof. The spaces with various heights and proportions allow flexibility in exhibiting art. Aluminum grid hanging ceiling is chosen to conceal structure, MEP equipment and integrate natural light and artificial light into a uniformed materiality to achieve a clean, consistent and neutral atmosphere optimized for showing art.

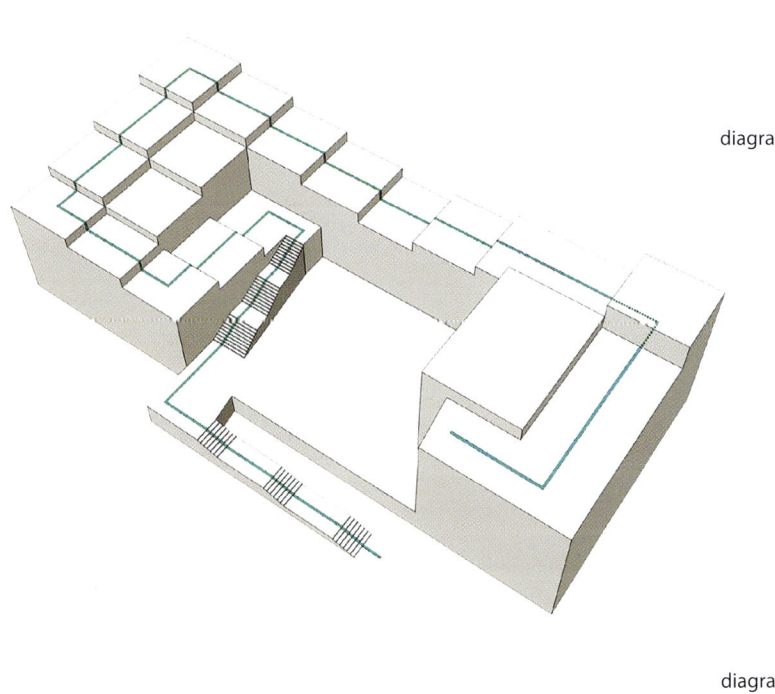

diagram-1

diagram-2

012

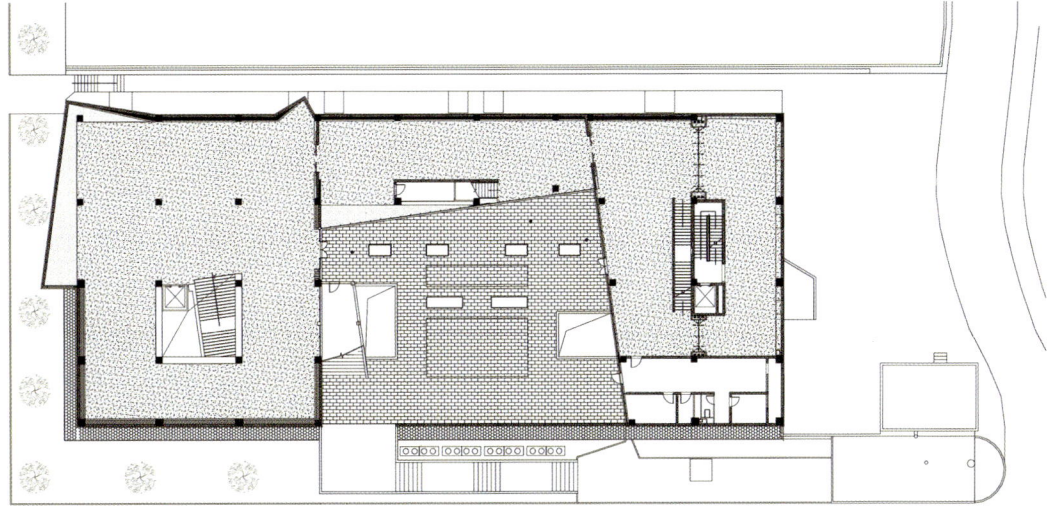

To receive outside views from inside was one of the aspirations of the design. The views are attentively revealed through a few precious protruding windows in the major exhibition spaces, making room for one to contemplate on the sight of reality before returning to art. The exterior wall is applied with wall tile that is economical and readily available in the market, the tiling style, however, shows a refined materiality that is in contrast to the surroundings, and further speaks of fluidity present in the spatial sequence of the architecture.

Somjai House

Architects: NPDA studio
Location: First Western Hospital - Koh Phangan, koh phangan, Thong Sala, Surat Thani 84280, Thailand
Architect in Charge: NPDA studio | Nutthawut Piriyaprakob
Area: 240.0 sqm   **Project Year:** 2015
Photographs: anotherspace
Interior Architect: NPDA studio | Nutthawut Piriyaprakob
Structural Engineer: Apisit Chawacharoen
System Engineer: Apisit Chawacharoen

The house was mainly designed as a home after retirement for the owner and, at the same time, it also serves as a reception for Coco-NutNume resort where family, friends and guests can stop by and have a good time.
The building consists of a main-bedroom with bathrooms, a guest-bedrooms, a guest-bathrooms and a pantry for family cooking. The main hall of the building serves as a flexible-multi-purpose space. It can be used as a living space, a party venue, coffee shop or even a yoga area. The upper deck was designed to provide panoramic view of the surrounded coconut plantation, the sea, Samui Island and the hills.

Sketch

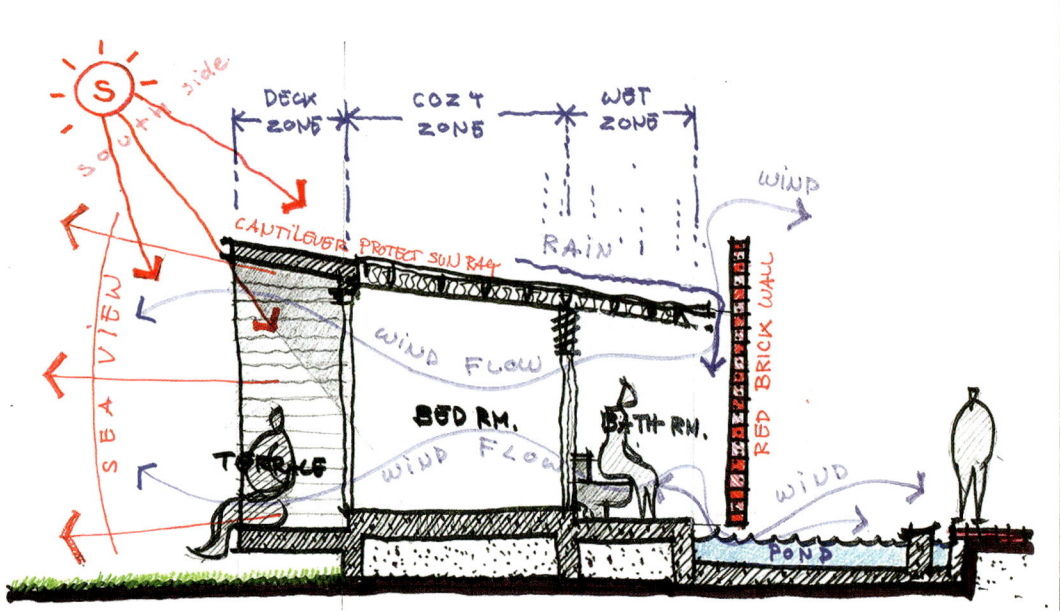

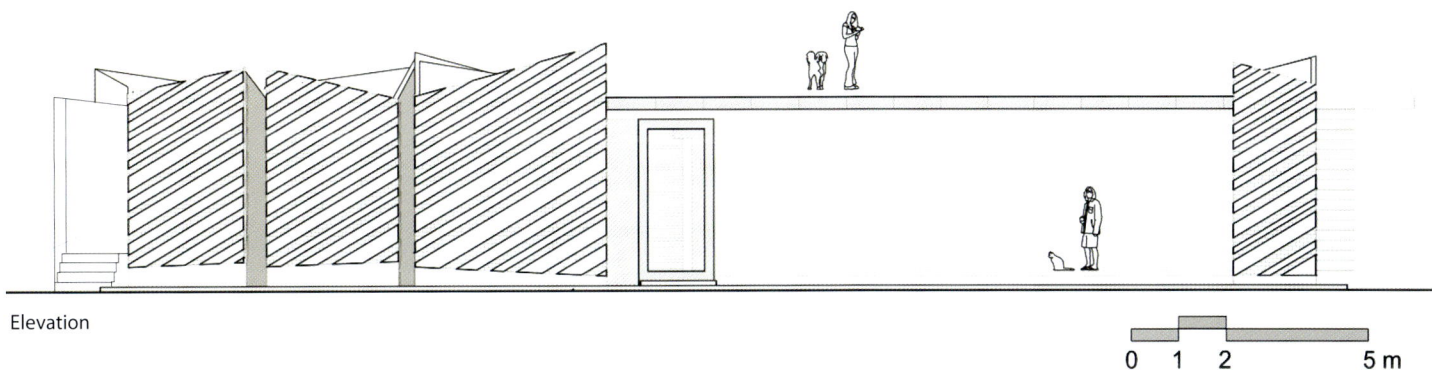

Elevation

0 1 2   5 m

Floor Plan

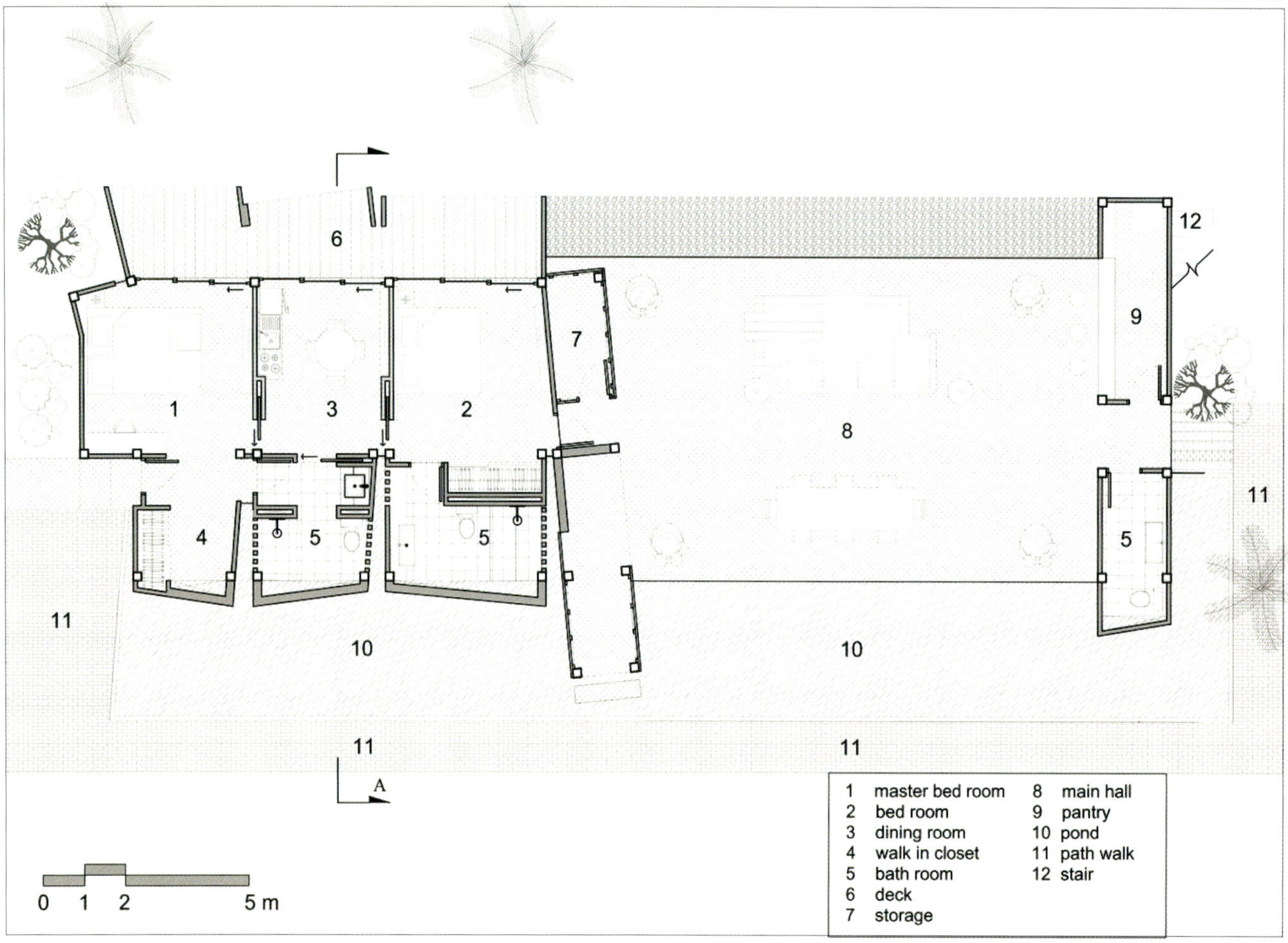

1 master bed room	8 main hall
2 bed room	9 pantry
3 dining room	10 pond
4 walk in closet	11 path walk
5 bath room	12 stair
6 deck	
7 storage	

017

The line of the architecture form is relating to the skyline of hills behind the building which allows the wind to flow smoothly with the curvy design. The front of the building was designed with open angle to receive the full view of the sea. The cantilever concept was used to protect the building from the hot sunlight.

The main color used was red to make the building stand out from the green of the coconut plantation and the blue of the sea. Brick and polished concrete were the material due the cooling property to protect the house from the heat and due to the local construction worker expertise. The thin iron staircase also shows the craftsmanship of local workmen as construction workers at the piers who are skillful in metal work.

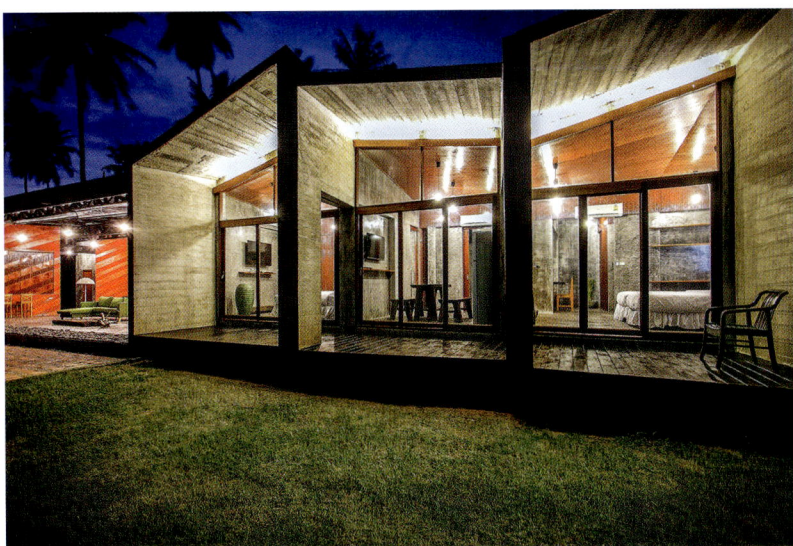

Toulkarem Courthouse

Architects: AAU ANASTAS
Location: Tulkarm
Funding: Foreign Affairs, Trade and Development Canada
Area: 8938.0 sqm   **Project Year:** 2015
Photographs: Mikaela Burstow

Toulkarem is a town in the northern part of Palestine, 15 Km west of Nablus, known to be a trading center for products from the city's surrounding villages and farms. Its geographical location in the country makes Toulkarem the most fertile city in Palestine.
Located at the north entrance of Toulkarem, the Courthouse takes part of a larger urban fabric remodeling.
The Toulkarem Courthouse houses the Magistrate and First Instance courts. The urban concept proposed, determines a succession of volumes. On Toulkarem's way to the city, a first building, at the top of the triangular parcel, accommodates public services while a second building unveils itself along with a locally cultivated interstice space.

West Elevation

020

Axonometric

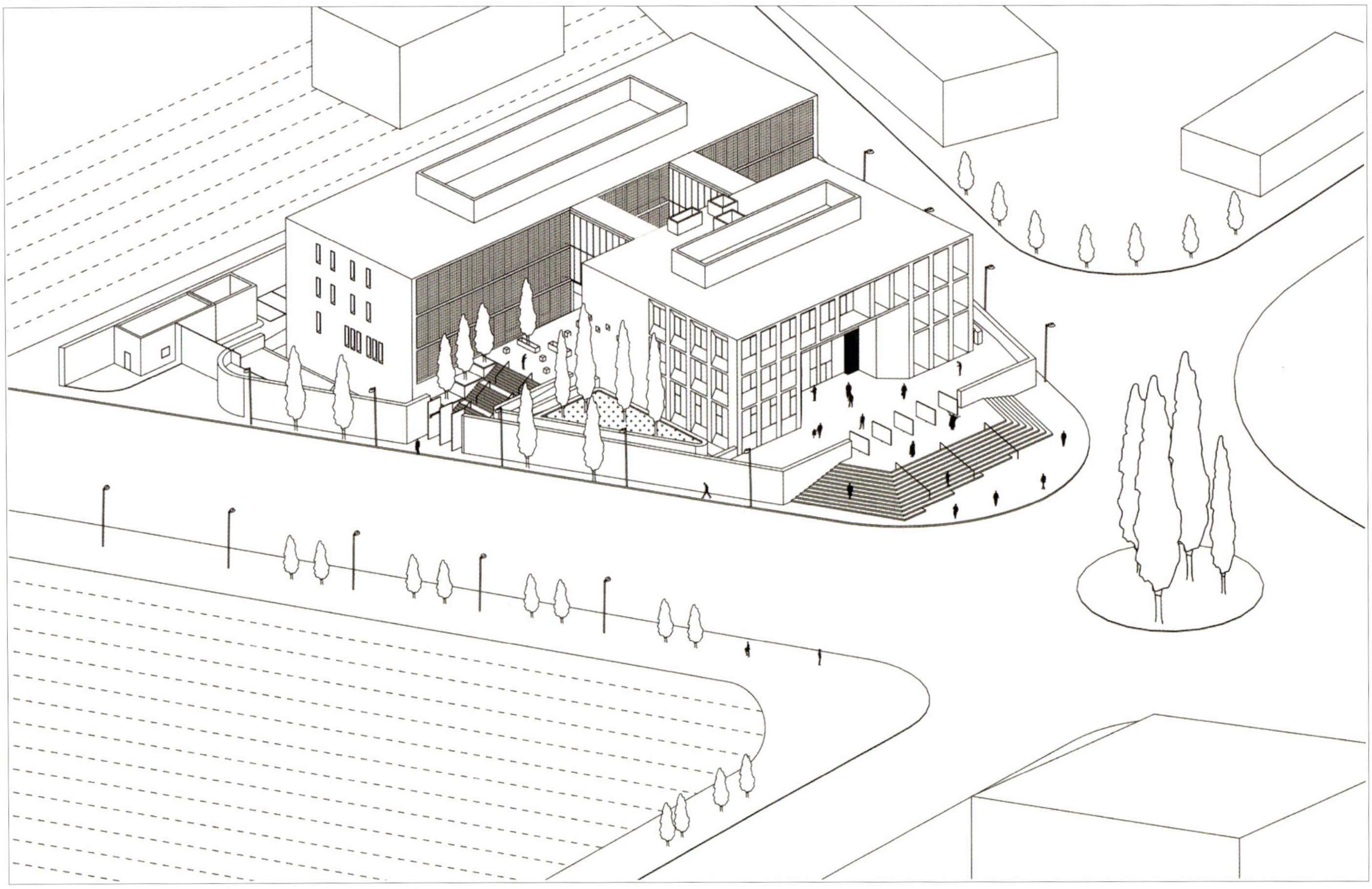

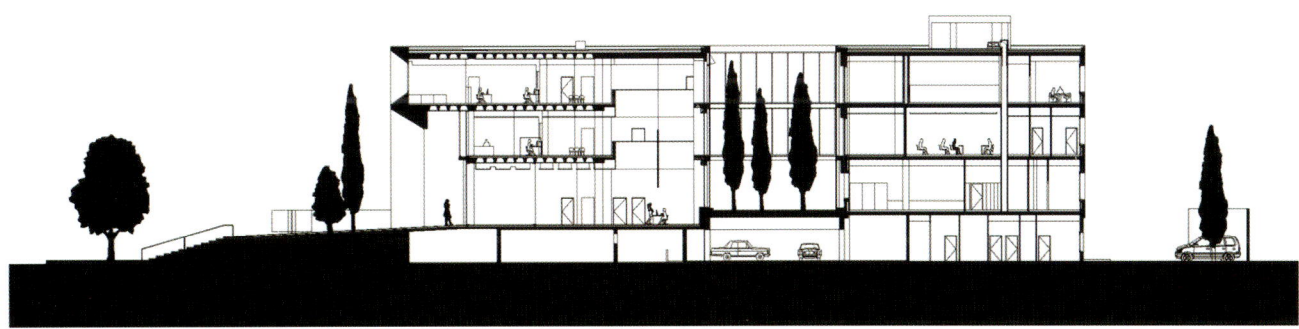

Cross Section

The main building is directly connected to a front public space, that anchors the building in its direct urban context and offers to the citizens of Toulkarem a social gathering space. The difference in levels between the street and the ground floor of the building, creates a public space setup that stimulates the citizens to invest the space.

The massive local stone walls of the main building are carved with different inclined geometries protecting the interior space from greenhouse effect during working hours, while offering large framed views of the neighboring landscapes. While entering Tulkarem, a public space introduces the first massive stone-carved-facade building after which a longitudinal planted interstice crosses the parcel and marks the shimmering steel geometric-patterned facade of the second building.

The succession of sequences is designed as an equilibrium of space allocations, architecture typologies and sun protection strategies. As a result, public space (in the sense of man-crafted space) contrasts with the Nature planted space, stone construction and light steel moucharabiyeh are put into tension, and claire-obscure effects are opposed to sun-reflective filter devices.

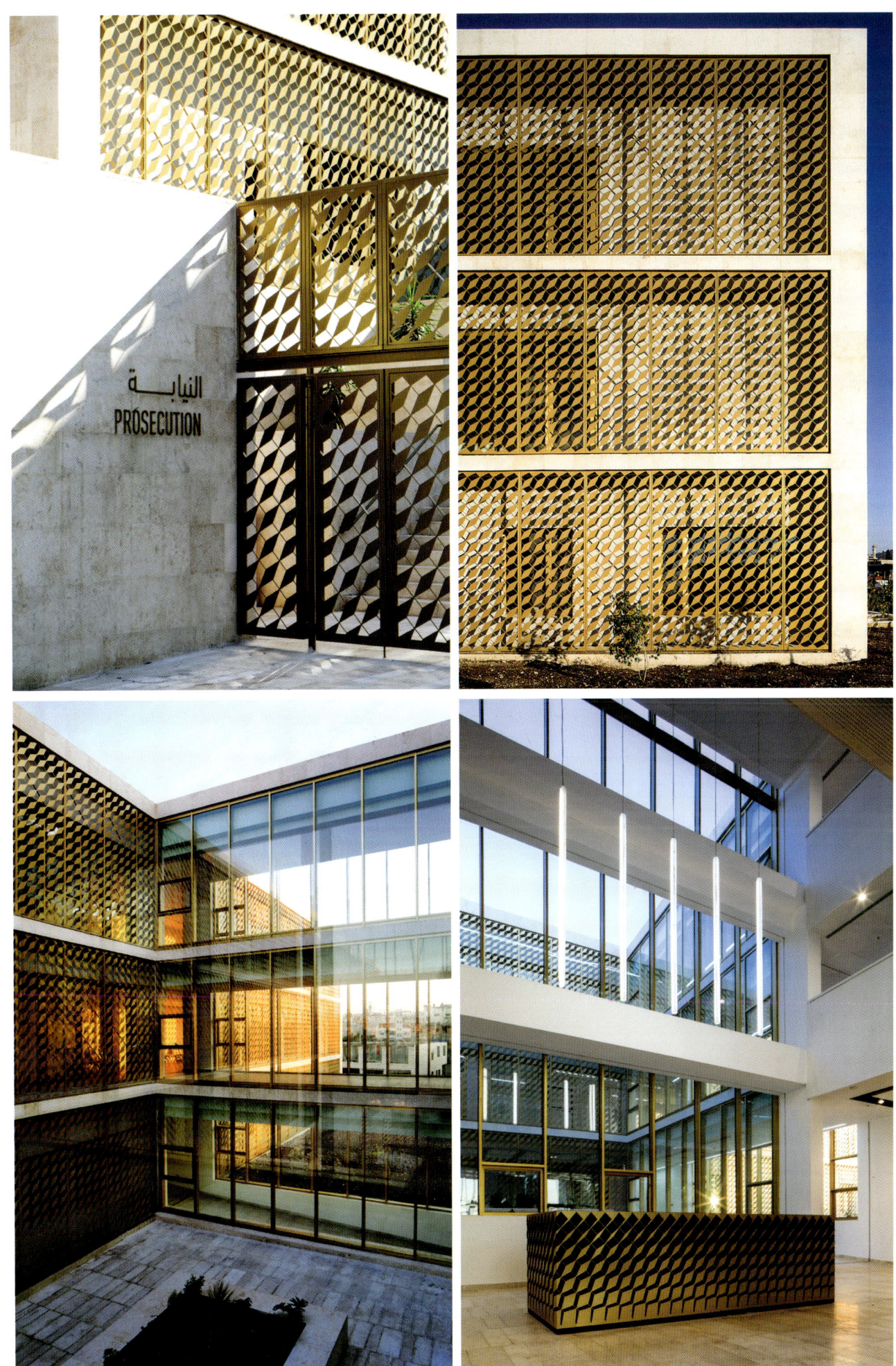

Mercado de Getafe Cultural Center

Architects: A-cero
Location: Getafe, Madrid, Spain
Architect in Charge: Joaquín Torres, Rafael Llamazares
Construction: Taller de construcción TMR
Project Year: 2015   **Photographs:** Rubén P. Bescós, Inés Mollá for RecordUs

A-cero presents one of his latest works. In this case, it is a public building which holds a cultural and leisure center for Getafe village, in the outskirts of Madrid. The main idea of the project is to reconvert completely a deserted building from years, which in a moment was the municipal market. The plot where is located, it is in the center of the city in the main square where it is also the city council. The status of the existing building was semiruins and the proposal of the city council was to give to the townspeople a multifunctional center focus above all in cultural events.

The plot has a rectangular shape with 510 sqm surface. The whole was composed for 2 buildings and an empty plot. The first one with two floors accommodates the main entrance ant it was the façade of "Plaza de la Constitución" (Constitution Square). The second building constituted the main part of the old municipal market; it was a diaphanous space with a big height with a lattice of trusses, which are cable –stayed with steel cables and together with the small plot, comprised the front of " Calle de los Jardines" (Garden Street)

The refurbishment project of the municipal market has 2 floors with a basement, having 750 of BUA. The distribution of the first floor has the hall, a big diaphanous all-purposes space, storage and services area. In second floor have 2 large lecture classes and common areas. In basement has all the storage area and MEP areas. Also a part of the stair and the lift has one service lift which communicates the 3 levels. In ground floor, with a useful area of 450sqm have the reception hall, wardrobe, toilets, the exhibition and activities room and access for goods.

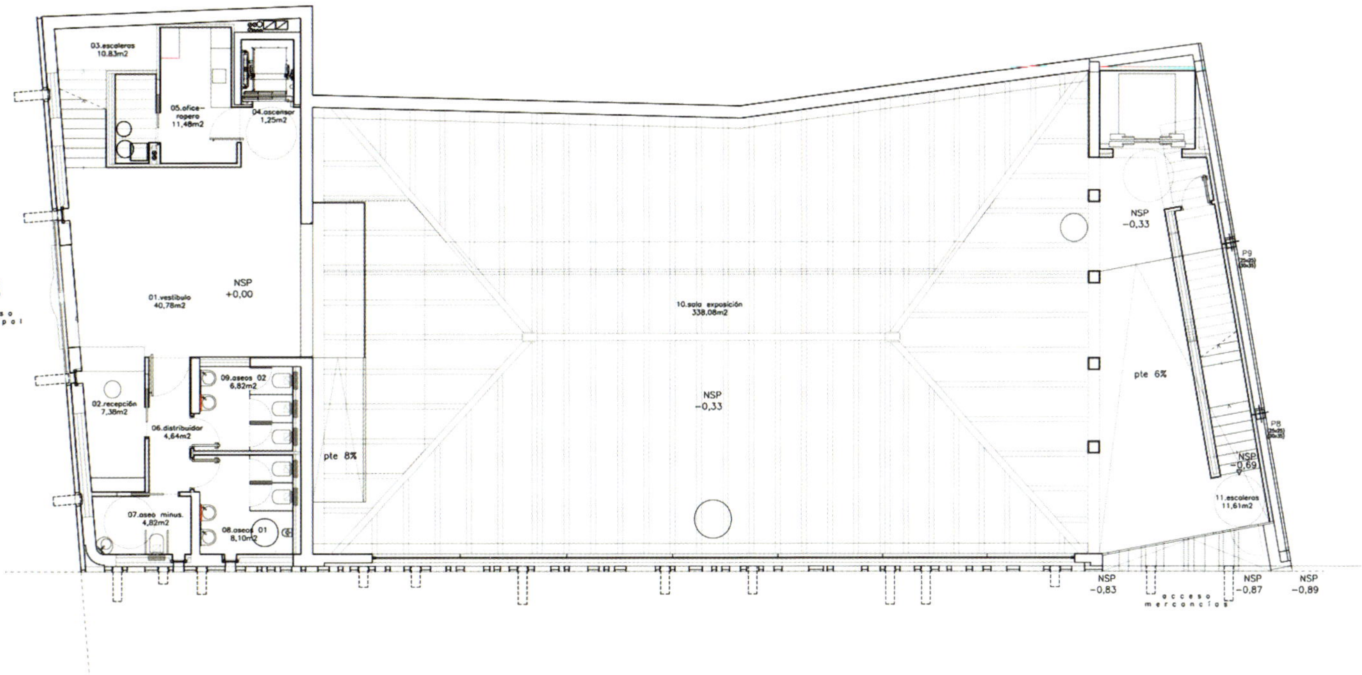

cuadro de superficies	
planta baja	
interior	**útil**
01. vestíbulo	40,78m2
02. recepción	7,38m2
03. escaleras	10,83m2
04. ascensor	1,25m2
05. oficio-ropero	11,48m2
06. distribuidor aseos	4,64m2
07. aseo minusválidos	4,82m2
08. aseos 01	8,10m2
09. aseos 02	6,82m2
10. sala de exposición	338,08m2
11. escaleras	11,61m2
total útil	445,79m2
total construida	497,03m2

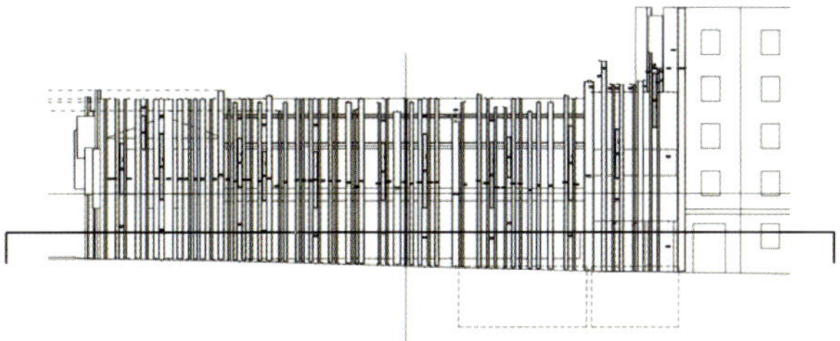

floor

026

It is proposed a modern image and technological which accommodate and protected the inside of the old building and preserve all the original character keeping the walls, the hollow ceramic ceilings, the concrete trusses … they will be enhanced using sustainable materials of the same industrial character of the building as the polish concrete, micro concrete, etc.

Outside the façade it has been designed as a skin done by white aluminum ribs which go across all the building, emulating the same rhythm of the interior trusses of the diaphanous space. This formal rhythm is broken with some pieces that goes out of the façade in different moments and that project pat on the external illumination of the building. From the façade you have the sense the interior of the big multipurpose area, coated by a second skin, which is the old brick façade that in the exterior it is painted in black. The façade which is located in the square, where it is the access to the main building the modern design it is interrupted with the conservation of the access of the old market ant the balcony.

Glass House at Sindhorn

Architects: OFFICE AT
Location: Bangkok, Thailand
Architects in Charge: Surachai Akekapobyotin
Area: 1650.0 sqm **Project Year:** 2015
Photographs: W workspace

Architect Team: Natthee Anuyotha, Chollatee Mayurarak
Structural Engineer: Sarawut Yuanteng
System Engineer: Mitr Technical Consultant Co.,Ltd.
Contractor: Thai Obayashi Corp., Ltd.

Sindhorn Building built in 1980, a more than 80,000 sq.m. Office building located on one of the most beautiful road in Bangkok called Wireless road, has been extended several times. Until now the owner has an intension to reimage the building by having a spectacular building in front of it to create new image and programs to support the building. Program: The main programs of the new building which is called "Glass house at Sindhorn building" are 4 fine dining restaurants with mezzanine floor and one small cigar bar. As well as, the site has a big tree that needs to preserve.

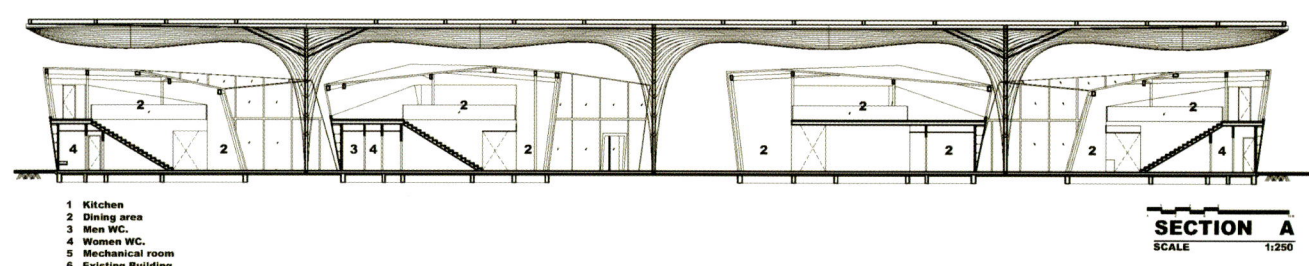

1 Kitchen
2 Dining area
3 Men WC.
4 Women WC.
5 Mechanical room
6 Existing Building

SECTION A
SCALE 1:250

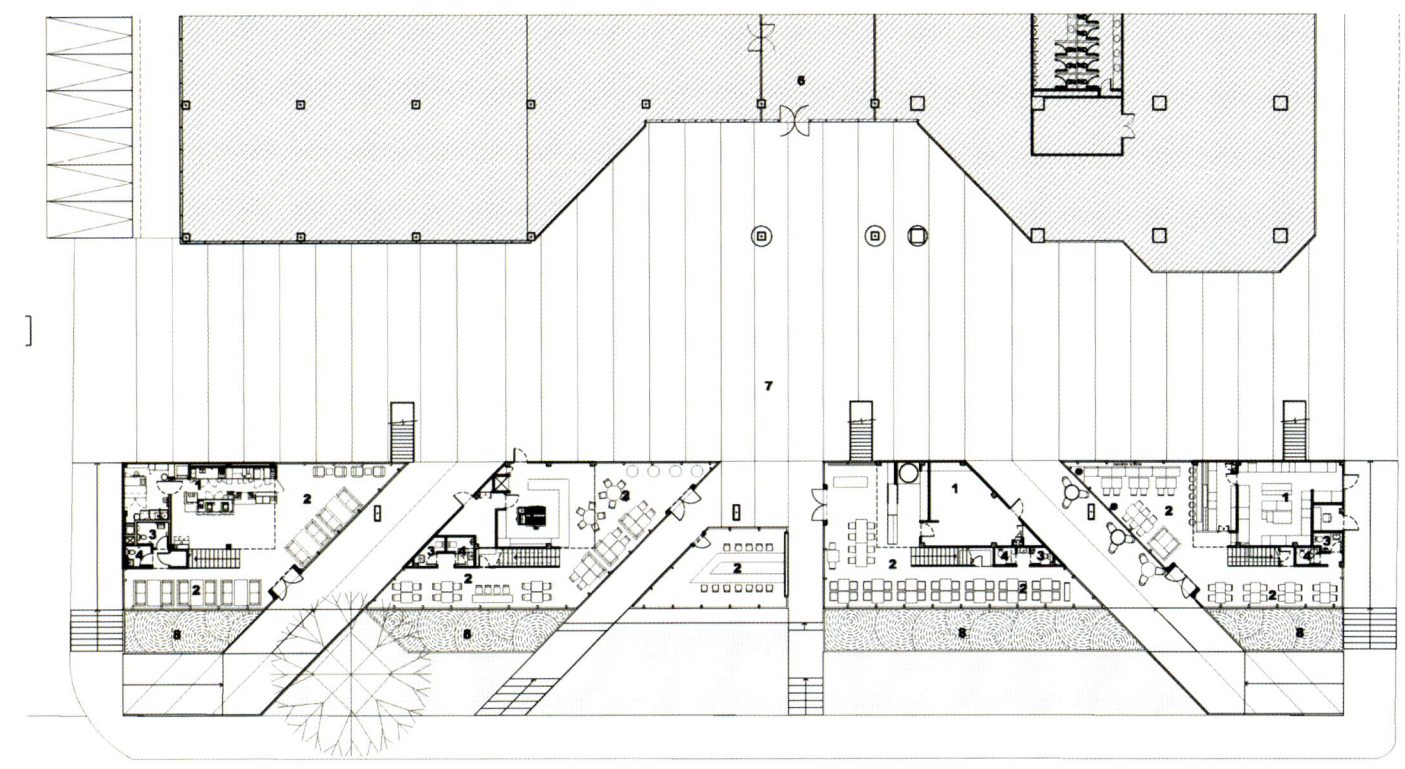

1 Kitchen
2 Dining area
3 Men WC.
4 Women WC.
5 Mechanical room
6 Existing building
7 Plaza
8 Reflecting pool

1st FLOOR PLAN
SCALE 1:250

032

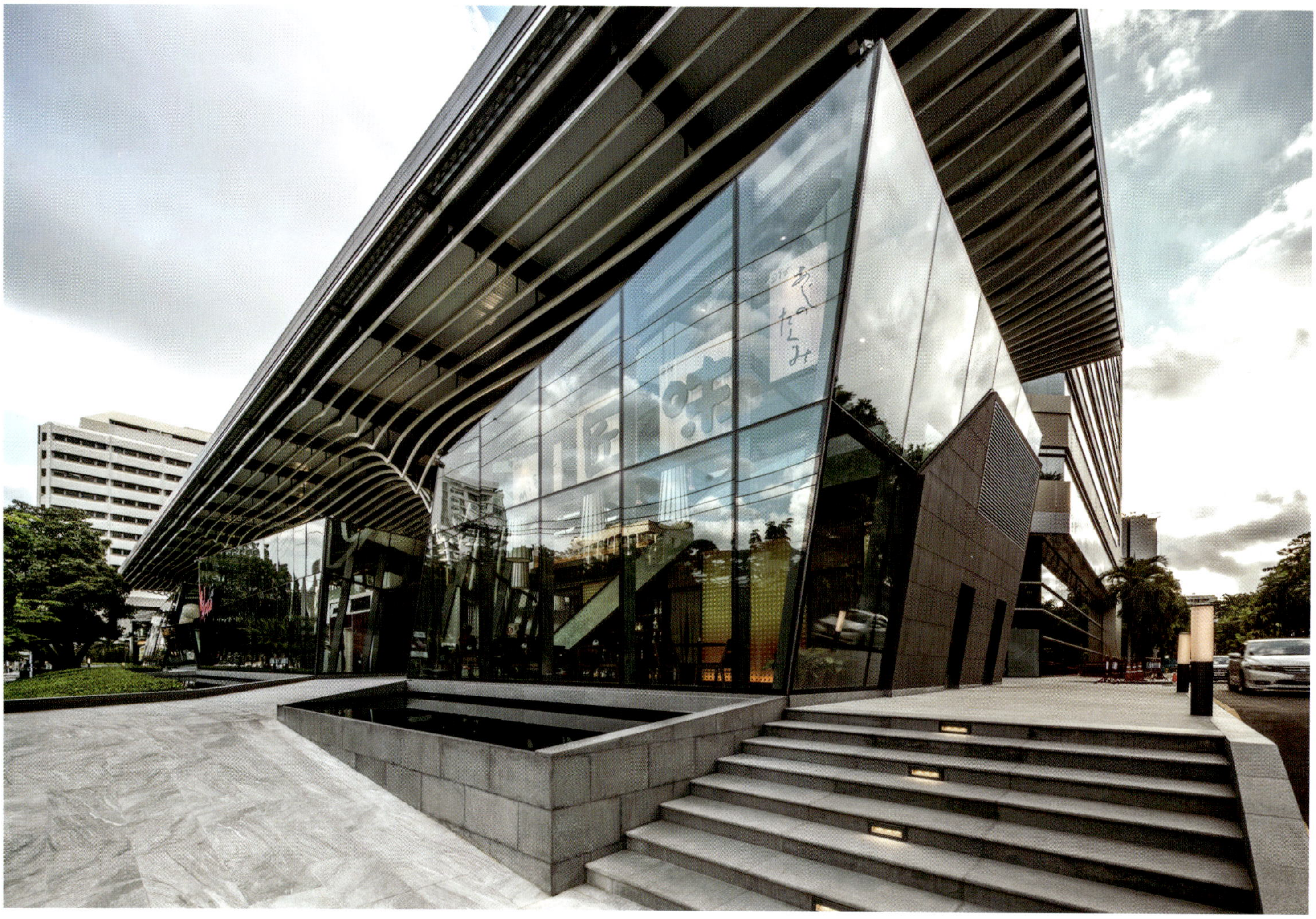

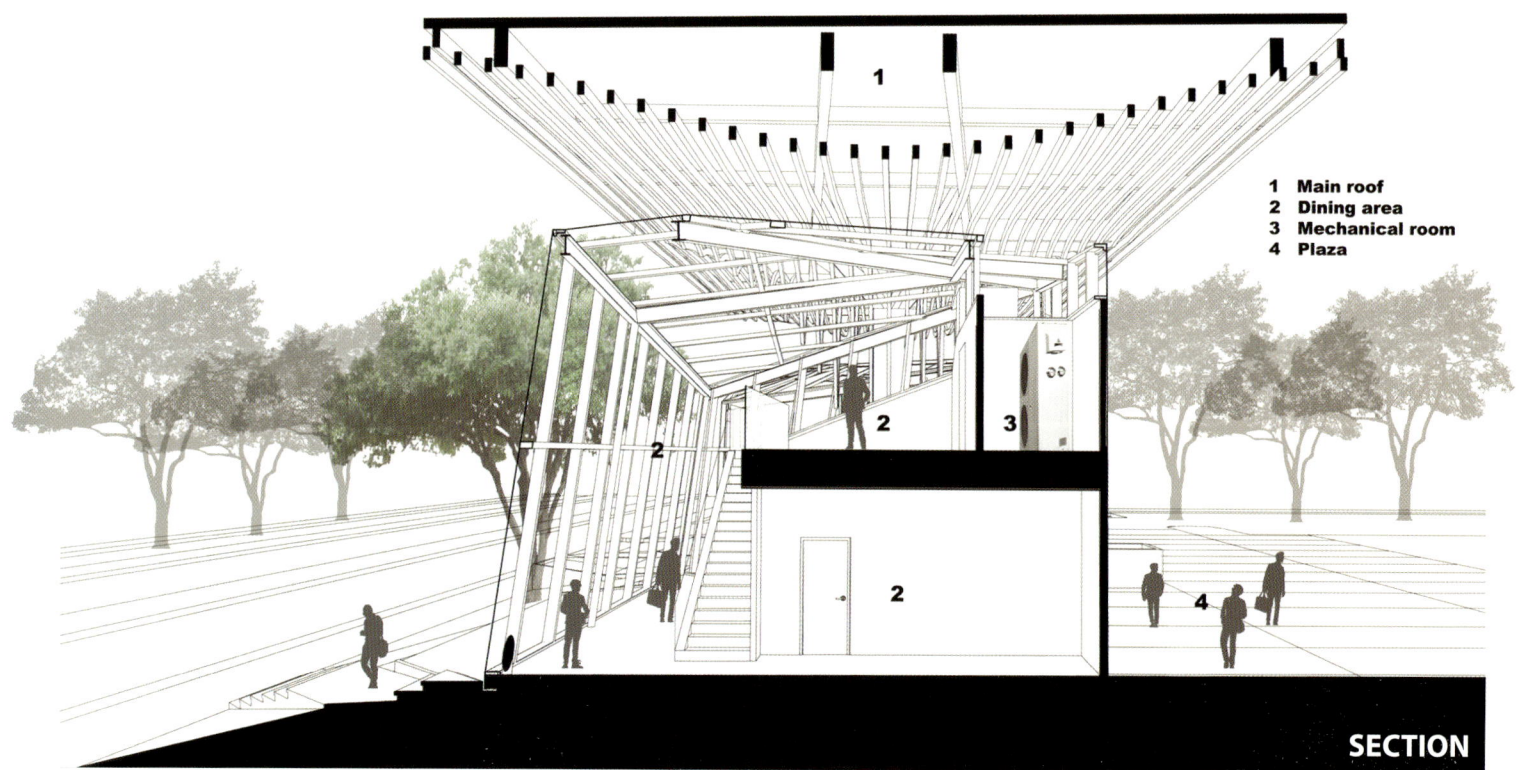

Separated: All fine dining restaurants are separated to connect indoor and outdoor space and create better dining atmosphere. Moreover, they are combined together with outdoor dining plaza.

1
2
2
3
2
2
4
1 Main roof
2 Dining area
3 Mechanical room
4 Plaza
SECTION

View & Ventilation: By splitting all restaurants apart, Each restaurant and Sindhorn building can easily access from public road. the new and existing buildings also get better view and ventilation too.

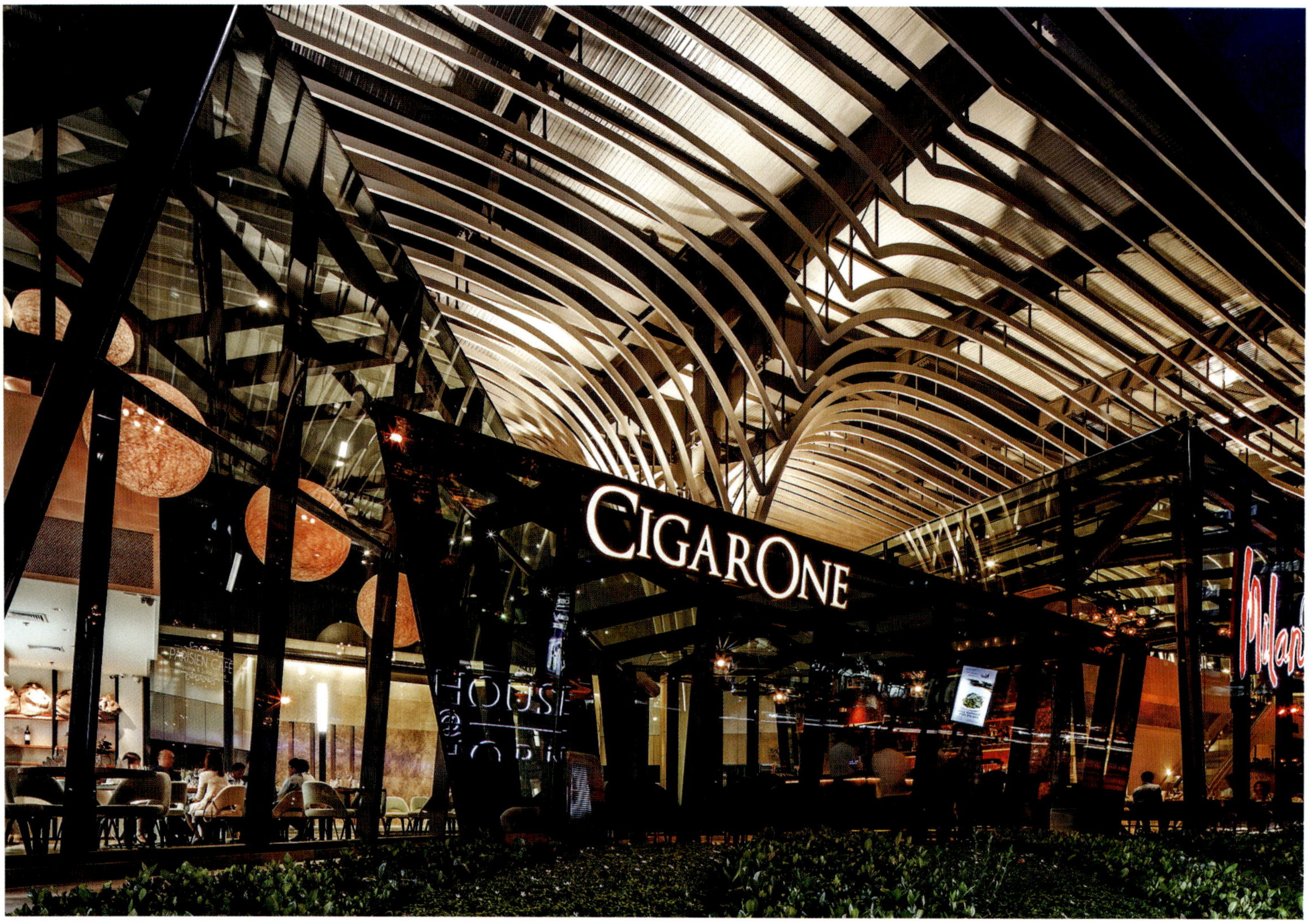

Icon: The new glasshouse, a brand-new icon of Sindhorn building, represents new image of the project, so it is shaped as crystal and this crystalized form is clad with glass to match its own concept.

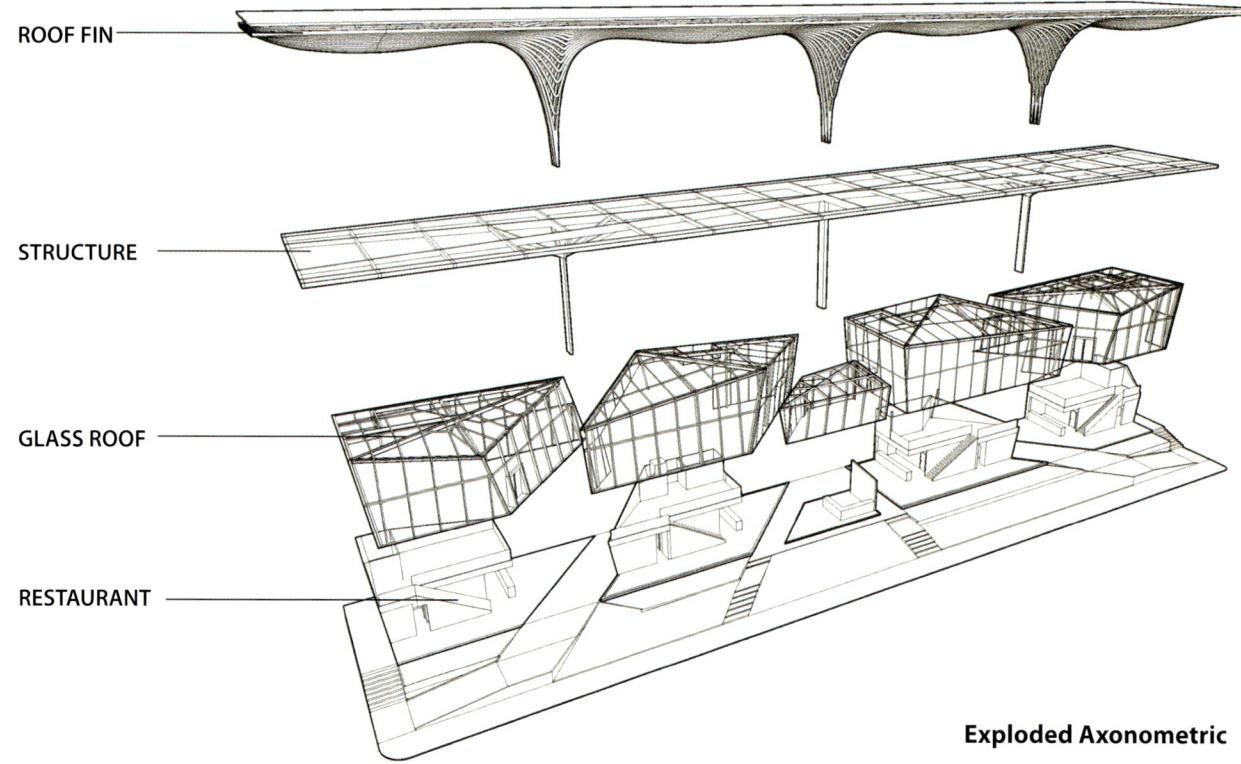

ROOF FIN

STRUCTURE

GLASS ROOF

RESTAURANT

Exploded Axonometric

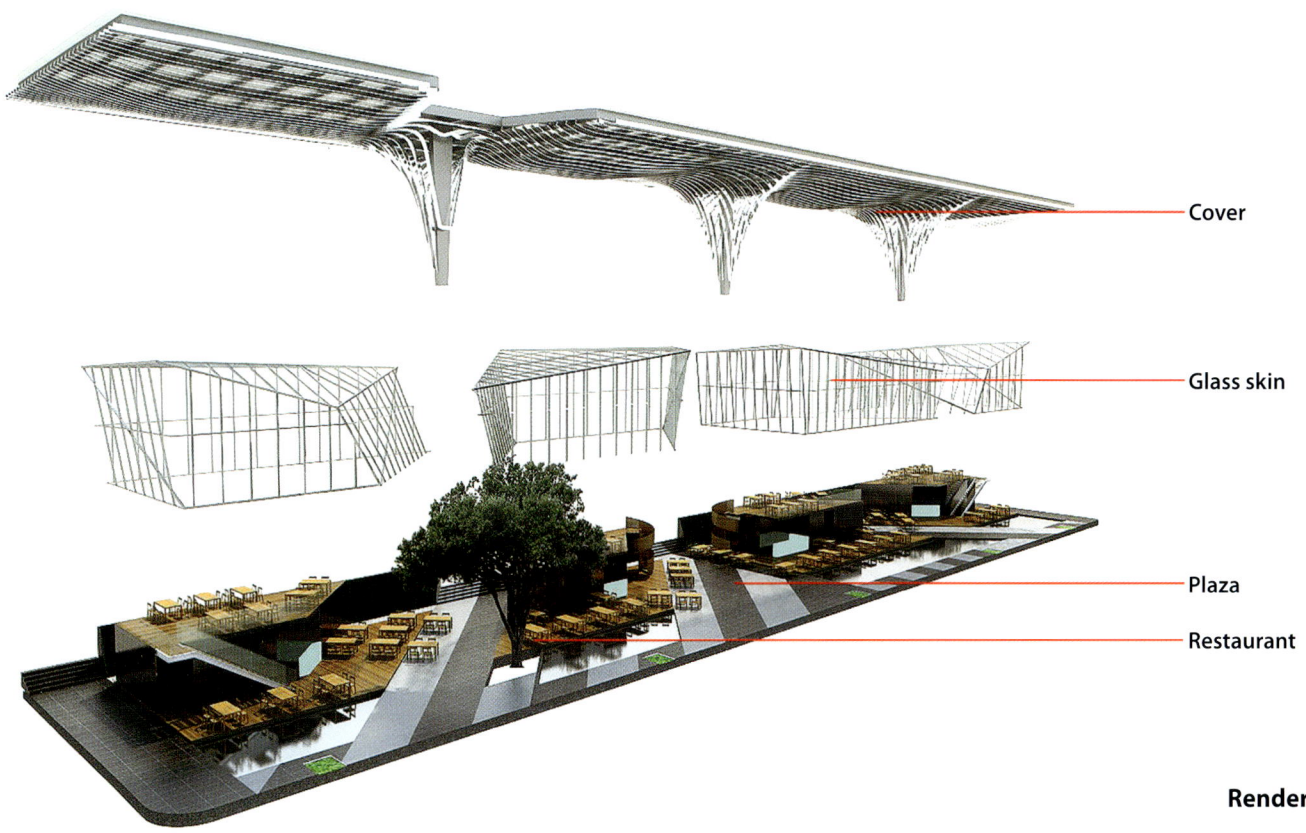

Shade: Moreover, the glasshouse location is on Wireless road where there are a lot of big trees as the road symbolic. The roof is designed specially to reflect the feeling of dining under tree shading.

Cover

Glass skin

Plaza

Restaurant

Render 2

036

Structure: The special structure takes double spans, each span length is 25 meter and at both ends are cantilevered as long as 20 meter, so that makes a 90 meter long roof. Moreover, this main structure is designed as the X shape to preserve the big tree in the site.

Branch: Additionally, to make the roof more appealing the roof is designed to have another layer of 31 horizontal steel fins that have irregularly shape as tree branches to create under tree shading mood and movement feeling underneath.

Diagram 2 - site

Diagram 3 - view ventilation

Diagram 4 - structure

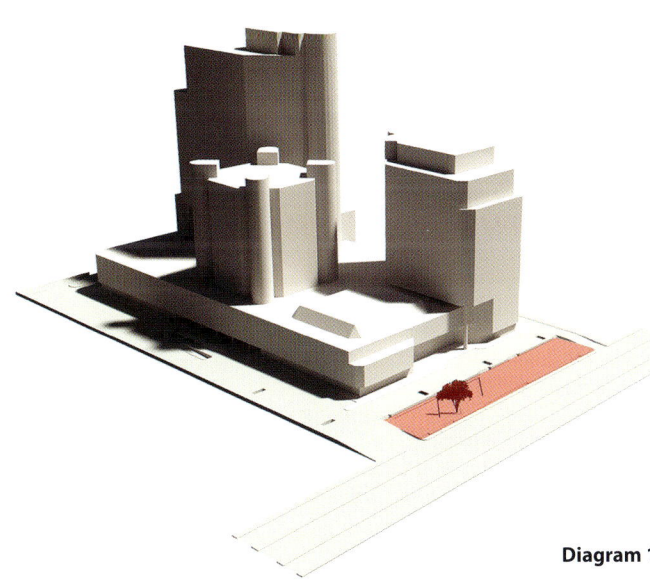

Diagram 1

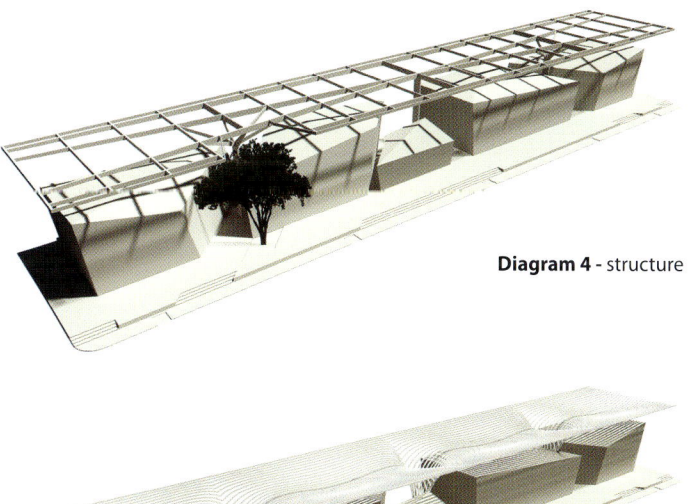

Diagram 5 - branch

BRANCH

038

Italy Pavilion

Architects: Nemesi
Location: Via Giorgio Stephenson, 107, 20157 Milano, Italy
Design: Nemesi & Partners Srl, Arch. Michele Molè Founder and Director and Arch. Susanna Tradati Partner Associate and Project Manager
Area: 27000.0 sqm   **Project Year:** 2015
Photographs: Courtesy of Nemesi

The design chosen for the Italy Pavilion is the result of an international design competition awarded by Expo 2015 SpA in May, 2013; among 68 participants Nemesi won the competition with Proger and BMS for the engineering and with Prof. Eng. Livio De Santoli for the sustainability.

The Italy Pavilion consists of the permanent building Palazzo Italia (6 levels, built area 14,398 sqm) and the temporary buildings along the Cardo (2 levels, built area: 12,551 sqm). Palazzo Italia reaches a height of 35 meters, the highest peak within the Expo site. It's the only permanently architecture at the Expo.

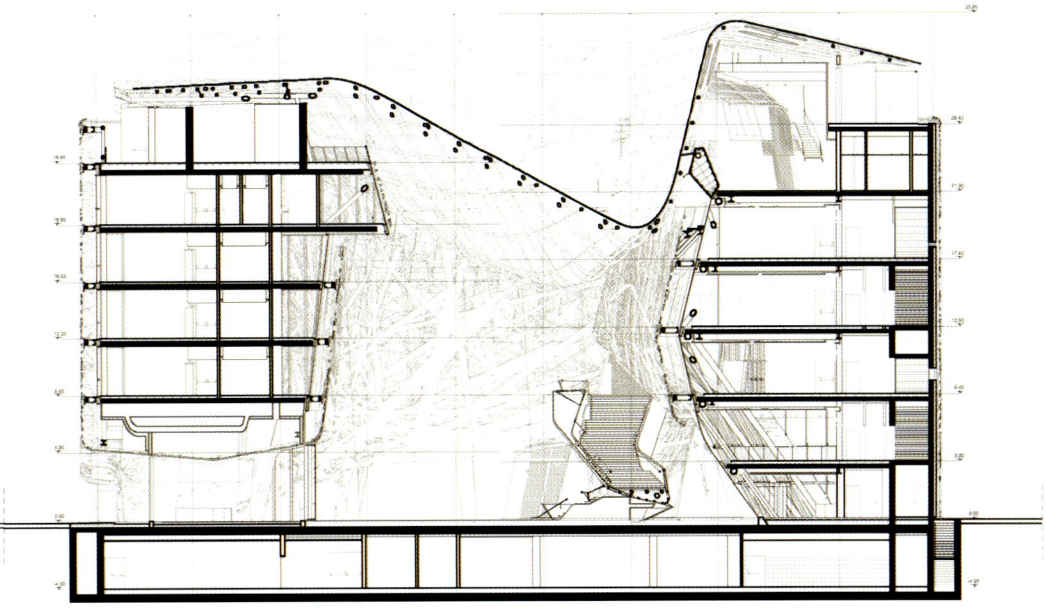

Section 1

SEZIONE AA'

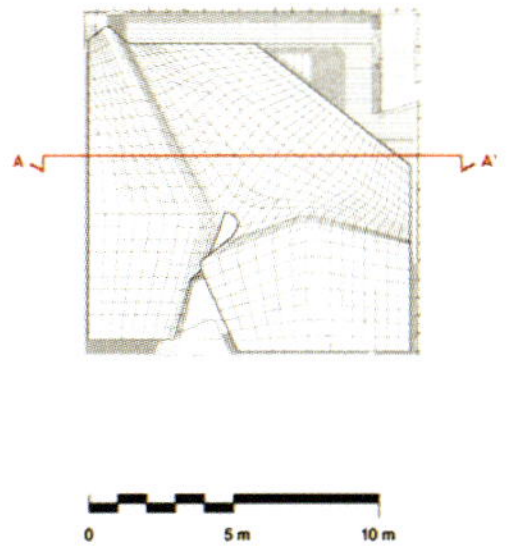

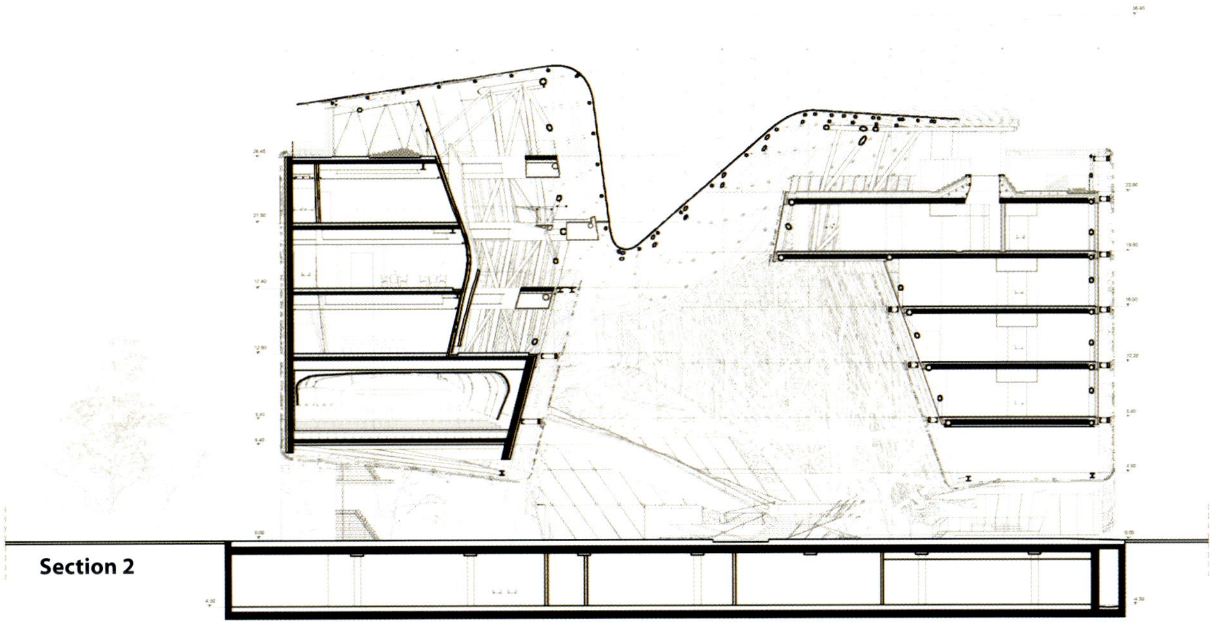

Section 2

SEZIONE BB'

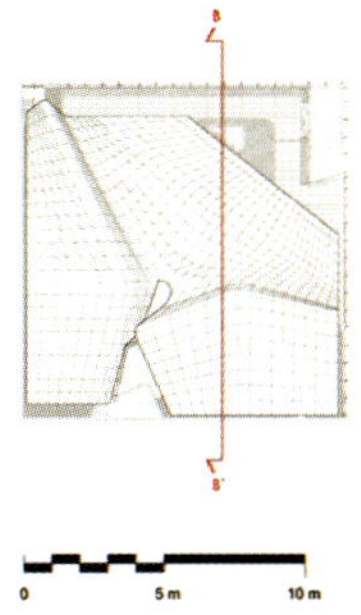

Palazzo Italia will host institutional spaces in addition to the excellences of "Made in Italy", while the Cardo temporary buildings will be representative of the Italian territory, in particular of the regions, and include a pavilion for the European Union placed in front of Palazzo Italia. Palazzo Italia, permanent building 60X60X34 mt (including branched façade and sail covering), includes: exhibition spaces, auditorium, delegations spaces, offices, events spaces, meeting spaces, restaurant. The Cardo temporary buildings contain: exhibition spaces, events spaces, offices, restaurant spaces and terraces events. Palazzo Italia is considered an architectural and constructive challenge for the complexity and innovation in design, materials and technologies used. The building is designed in a sustainable way thanks to the contribution of photovoltaic glass in the roof and the photocatalytic properties of the new concrete for the branched facade. 2,000 tons of i.active Biodynamic concrete over 700 branched panels all differents 4,000 sqm of sail covering - 400 tons of steel.

Because of its architecture and its location, Palazzo Italia is a Landmark within the Expo site: located to the north, it is the scenic backdrop for Cardo avenue which runs right across the site.
For Nemesi, the spark for Palazzo Italia was a concept of cohesion in which the force of attraction generates a rediscovered sense of community and belonging. The internal piazza represents the community's energy. This space - the symbolic heart of the complex - is the starting point for the exhibition route, in the midst of the four volumes that make up Palazzo Italia.

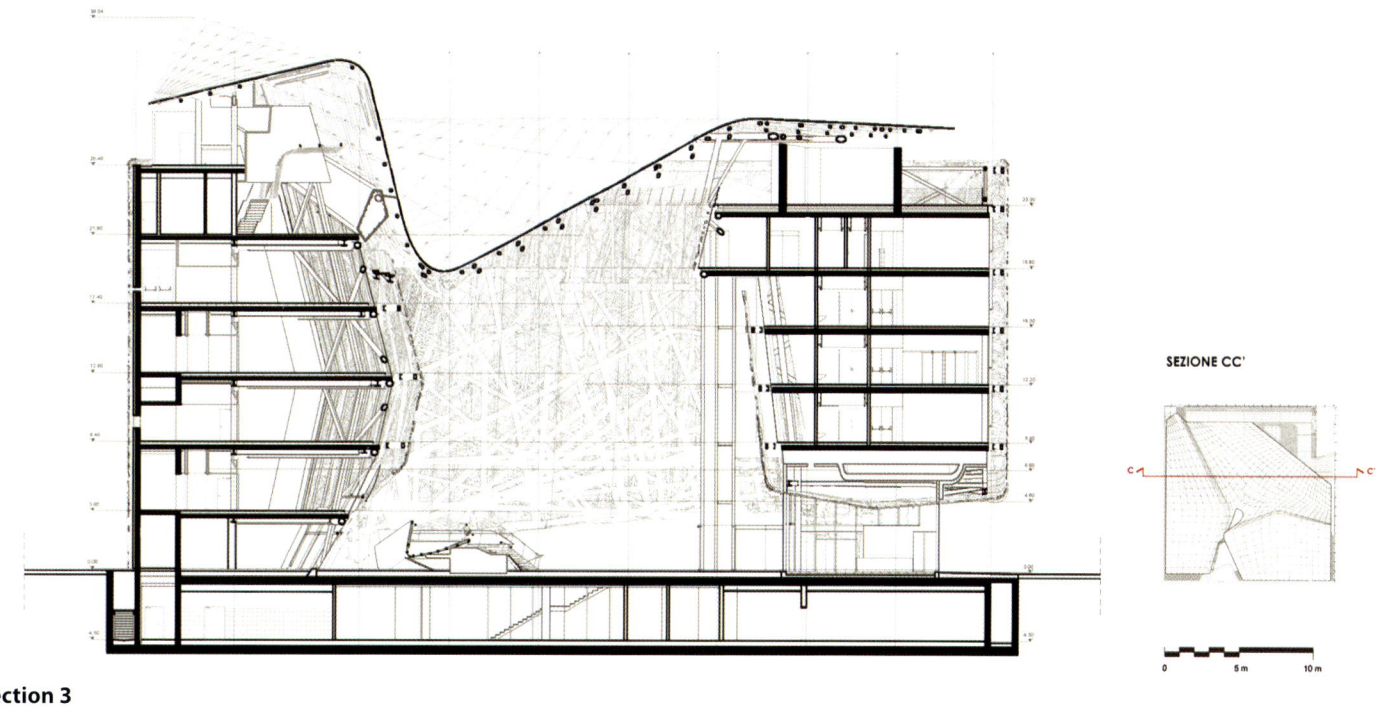

Section 3

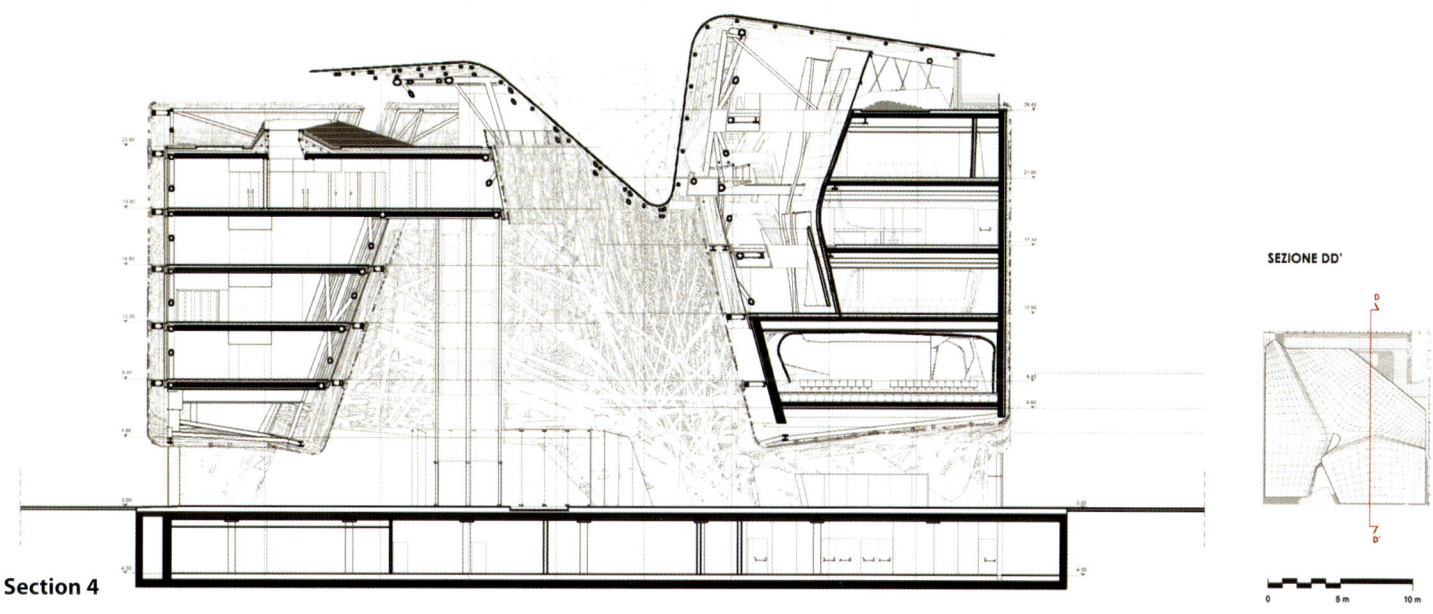

Section 4

Palazzo Italia draws on the concept of an "urban forest" with the branched outer envelope designed by Nemesi. For the design of this "skin" Nemesi has created a unique and original geometric texture that evokes the intertwining random branches. The full external façade of Palazzo Italia will be clad in over 700 i.active BIODYNAMIC panels realized by Styl-Comp with Italcementi's patented TX Active technology. When this material comes into contact with light, it can "capture" pollution in the air, transforming it into inert salts and reducing smog levels.

The roof designed by Nemesi for Palazzo Italia is an innovative "sail" realized by Stahlbau Pichler. It's an interpretation of a forest canopy, with photovoltaic glass and flat and curved geometric shapes (often squares). Together with the building's envelope of branches, it will be a manifest expression of innovation in design and technology. The roof reaches its architectural height above the inner piazza, where a massive glazed conical skylight "hangs" over the square and the central steps, radiating natural light.

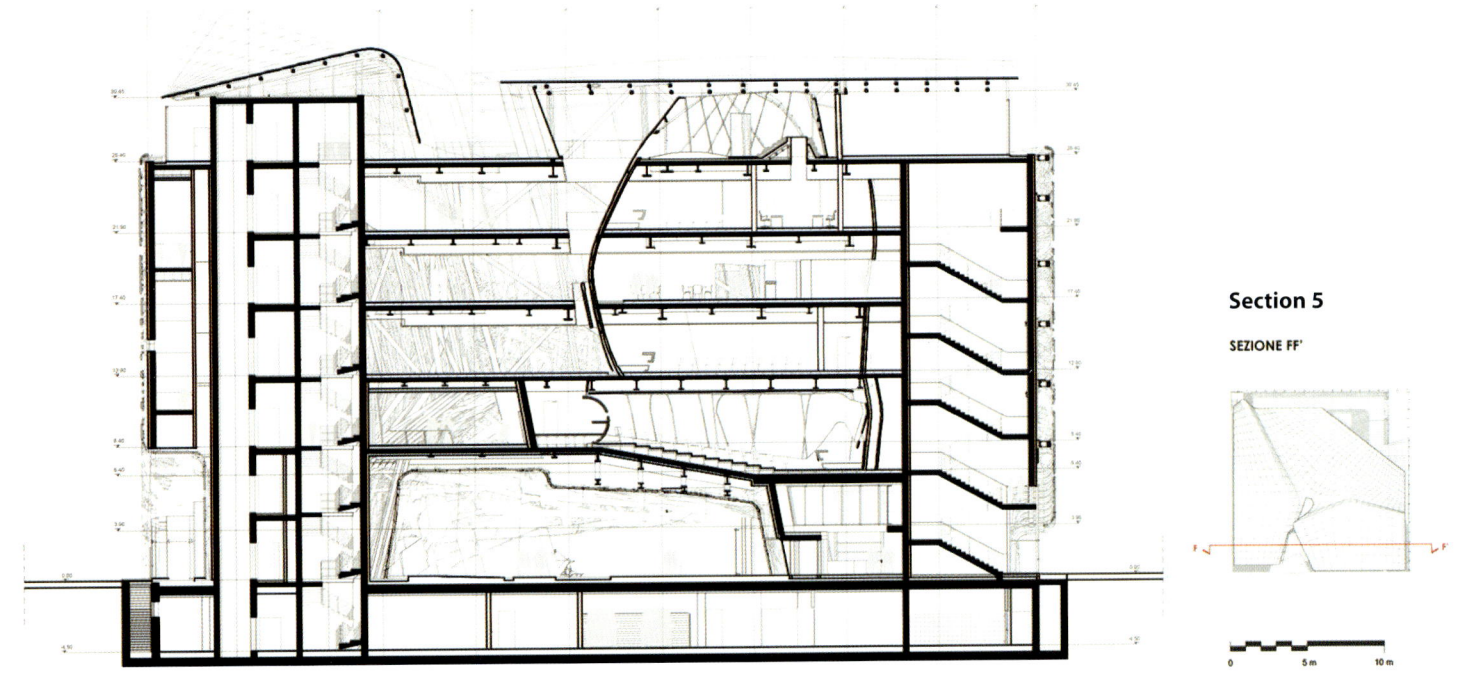

Section 5

SEZIONE FF'

044

Lima Duva Resort

Architects: IDIN Architects
Location: Ko Samet, Thailand
Architects in Charge: Jeravej Hongsakul, Sethapong Pisithwanish, Oak Sornnill, Jutiporn Taerattanachai
Interior Designer: Jureerat Korvanichakul
Project Year: 2015   **Photographs:** Spaceshift Studio

Lima Duva is a new phase of Lima Bella, located on Aow Praw, Koh Samed, Thailand. Koh Samed is a popular destination among modern couples who seek a place for romance, hence the phrase "Pai Samed Sed Took Raii," referring to the island's enchanting atmosphere. In Thailand's 19th century literature, "Pra Apai Manee," its revered poet Sunthorn Phu used Koh Samed as the setting for a passionate chapter. The design team took the notion that Koh Samed was generally perceived in relation to love and romance, thus the design development idea was to create a place that serves couples and their activities, while not alienating the family guests. Each unit has its semi-outdoor balcony acting as a divider from the adjacent unit to maximize the privacy.

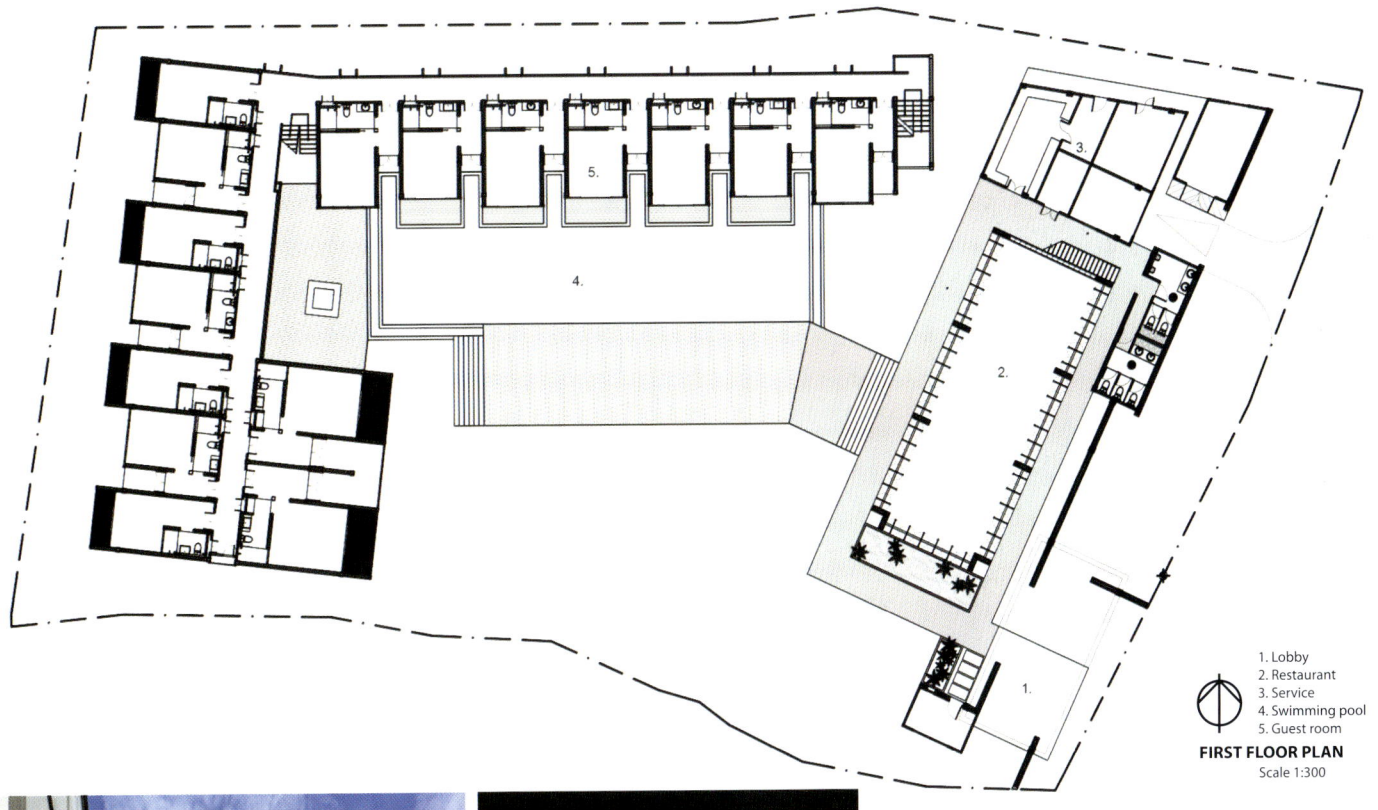

1. Lobby
2. Restaurant
3. Service
4. Swimming pool
5. Guest room

FIRST FLOOR PLAN
Scale 1:300

All ground floor units have access to the pool and a private Jacuzzi for two occupants. The color of the pool tiles gradually become darker at the Jacuzzi area to create visual privacy for the guests. Ventilation blocks are used along the corridors to bring in the sunlight in a pattern that varies throughout the day, creating different impressions when walking back and forth.

Elevation C
scale 1:250

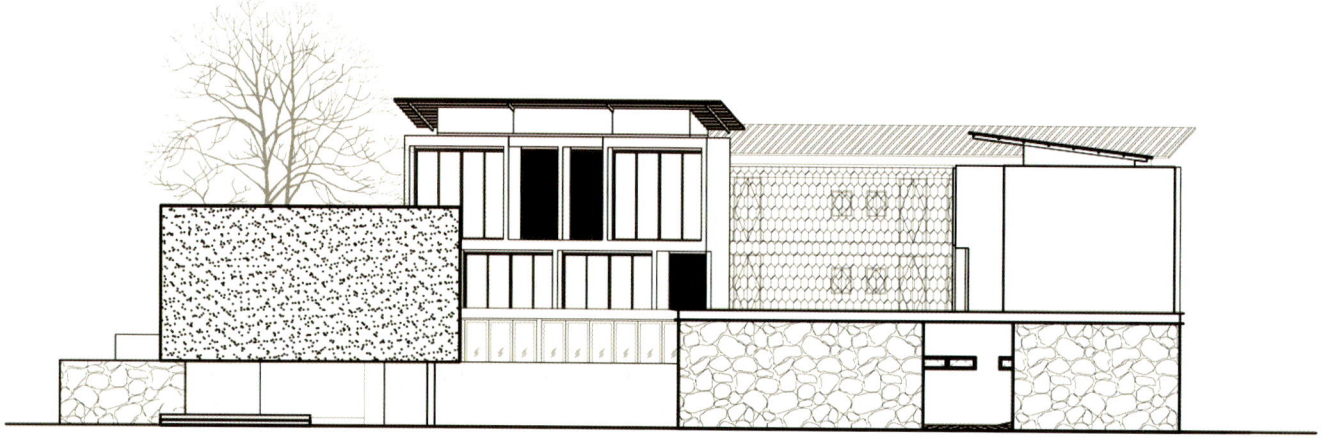

The lobby was designed in a simple white geometric form located close to the main road, enclosing the courtyard and the existing trees. At the entrance guests need to bend down slightly as the space leads them in, gently revealing parts of the resort. The white box lobby can also be used as a screen for films and other media projection, using the courtyard as the seating area.

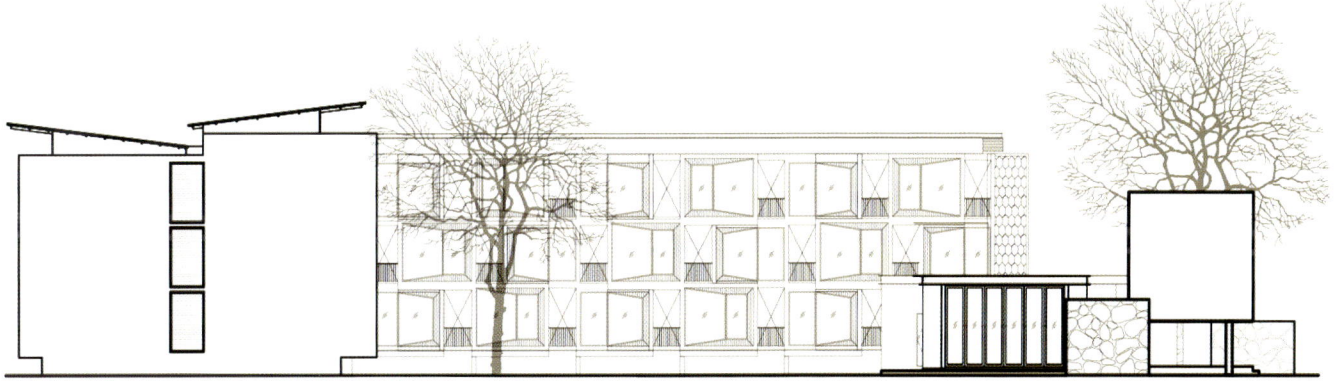

Elevation A
scale 1:300

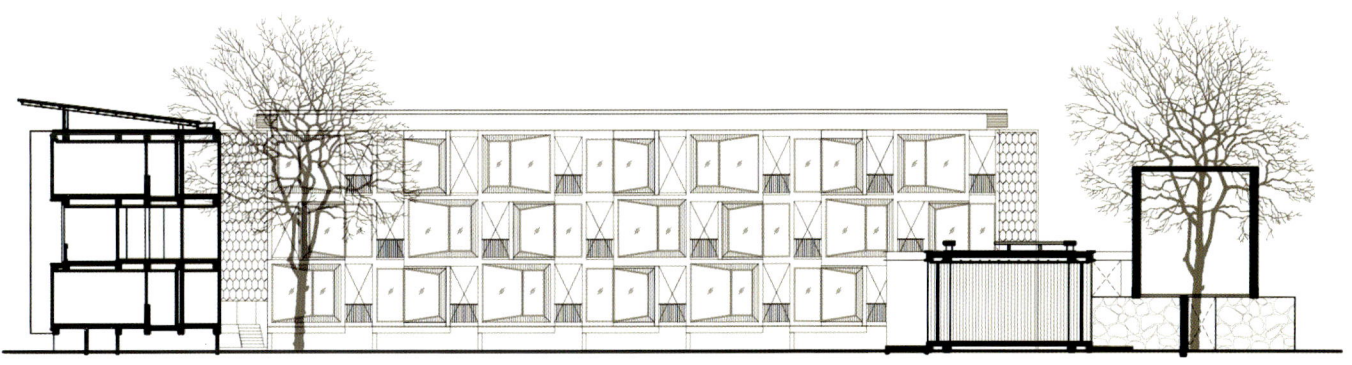

Elevation B
scale 1:300

Section A
scale 1:300

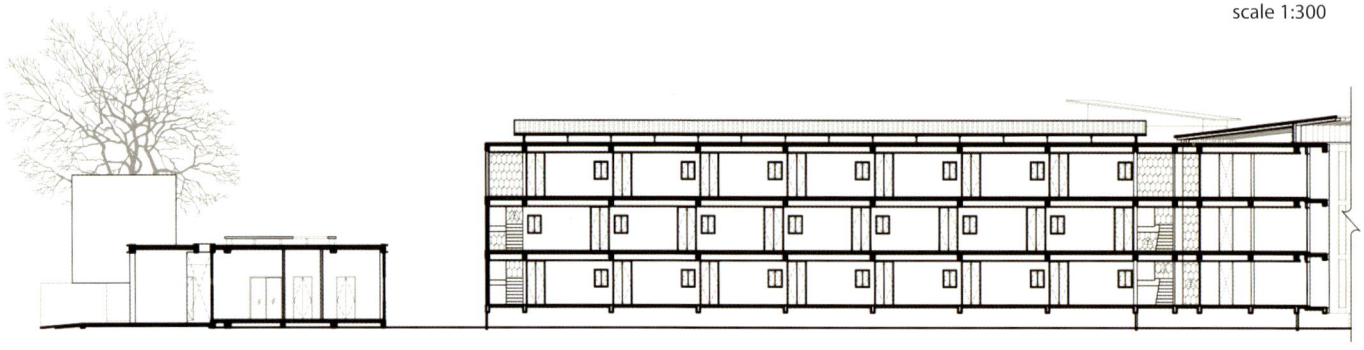

Jesuit High School Chapel of the North American Martyr

Architects: Hodgetts + Fung
Location: JESUIT HIGH SCHOOL, 1200 Jacob Lane, Carmichael, CA 95608, USA
Area: 10478.0 ft2   **Project Year:** 2014   **Photographs:** Joe Fletcher
Structural Engineer: Thorton Tomasetti   **Mechanical Engineer:** Capital Engineers Consulting, Inc.
Landscape: Yamasaki Landscape Architects   **General Contractor:** Swinerton Builders

Section

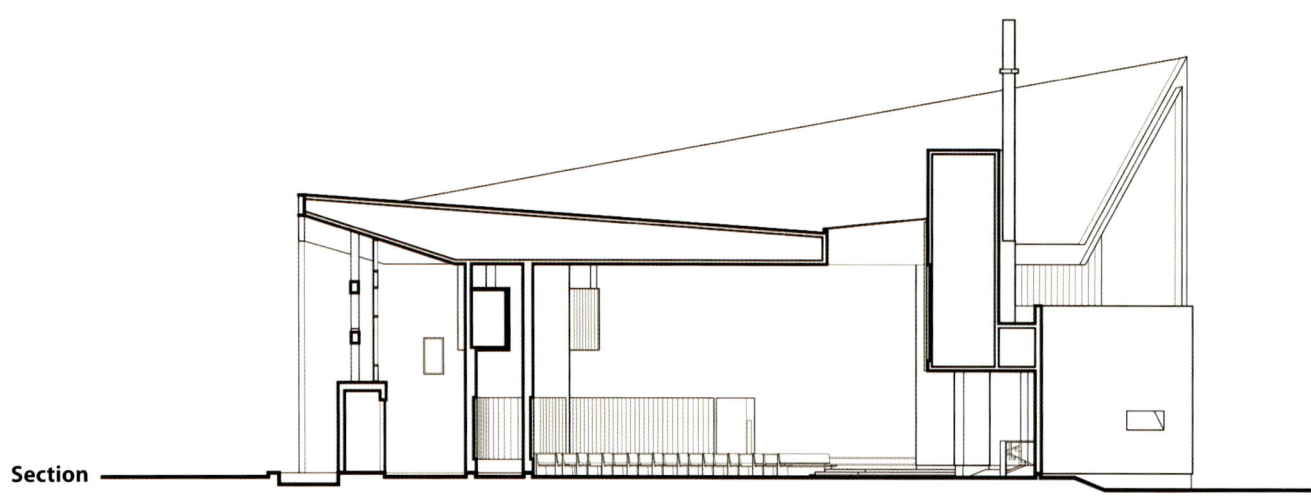

The design for the Jesuit Chapel is intended to create a contemplative journey for each student. This journey begins at the newly created entrance to the campus from Fair Oaks Boulevard. The gentle curves of the drives and pathways take their queue from the existing JHS campus. An extension of the existing campus pathways is created to embrace a plaza and a wooded field which envelop the space for the chapel.

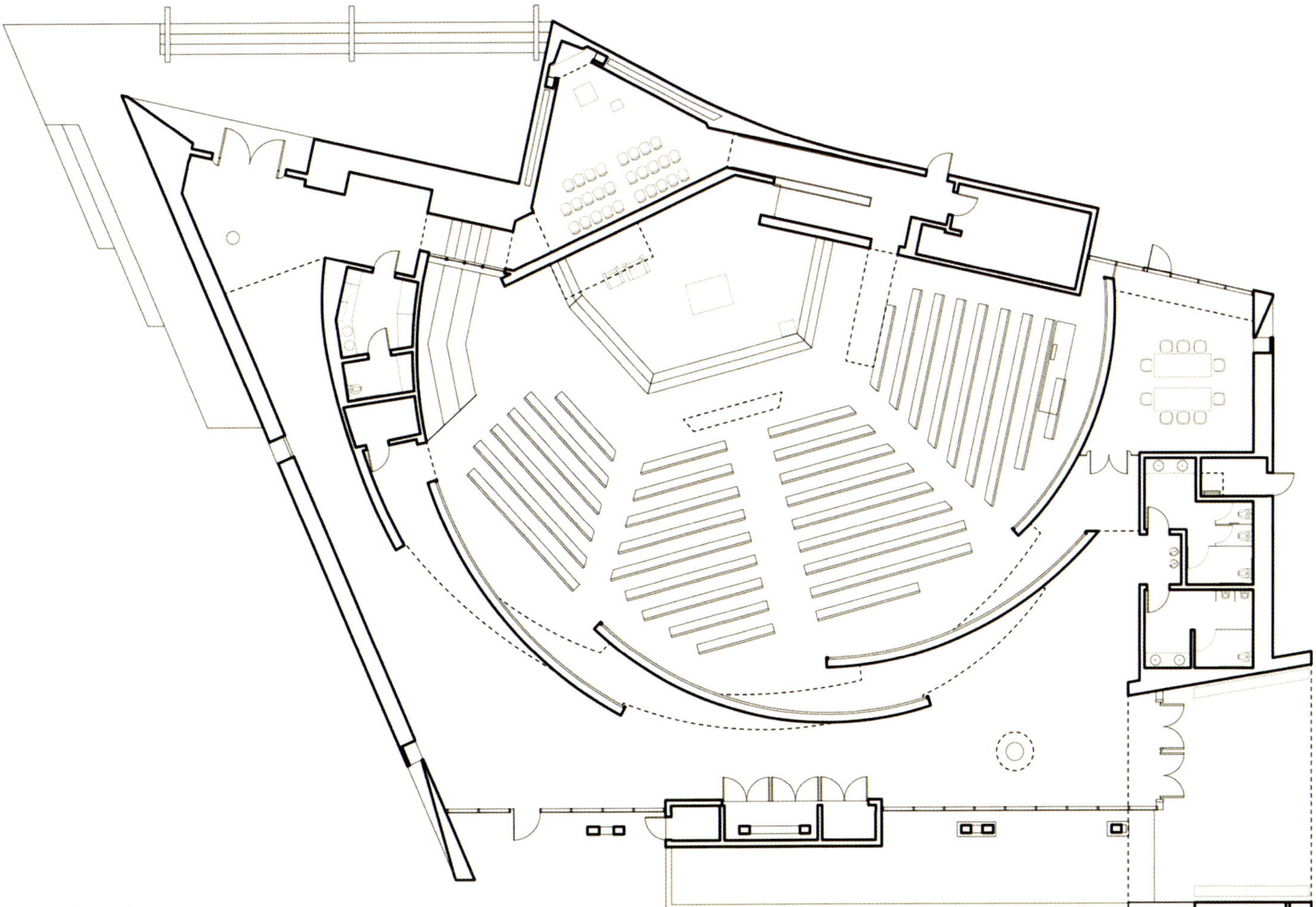

Floor Plan

The Chapel is positioned as the new apex for the Jesuit High School campus; in its position it becomes an integral piece to the overall master plan of the school while maintaining a visual presence to the general community. The placement and height of the chapel reflects its importance as the signature building for Jesuit High School. The orientation of the building directs the highest point of the roof towards the student arrival area creating a transition point for students entering the campus; the lowest points of the roof relate to the heights of the adjacent campus buildings.

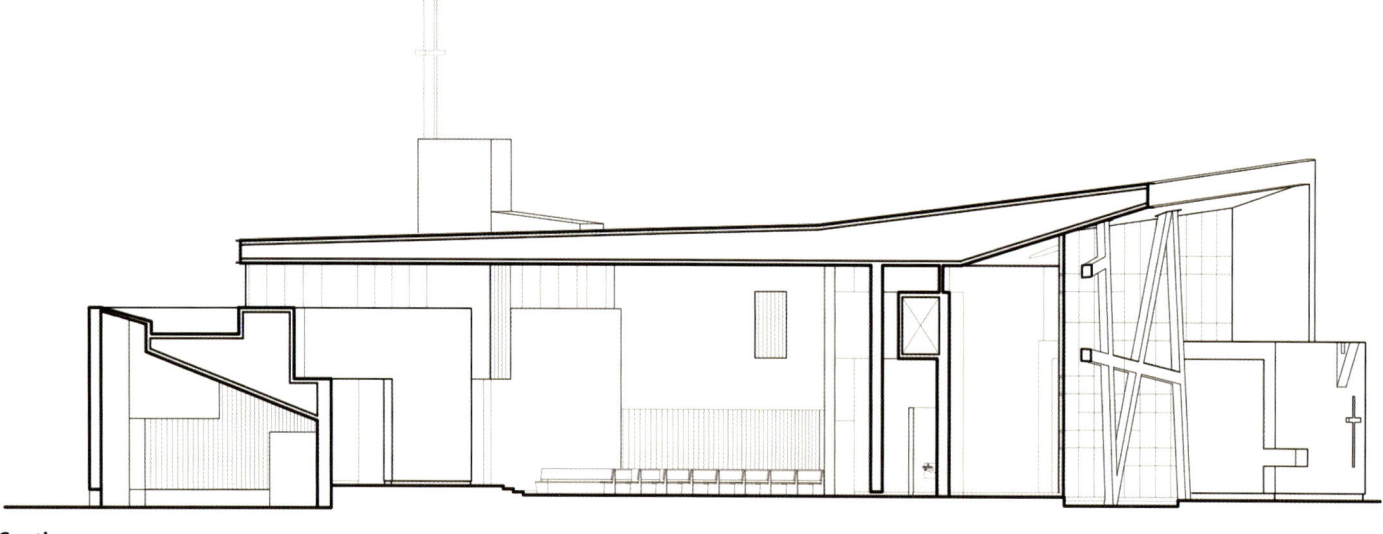

Section

The simple geometry and limited material palette of the chapel creates subtle entrances and gathering spaces for students. Entrances are defined by a change in materials and lowered ceilings. The student entrance is orientated along the east/west plaza connecting to the existing campus, and is the most prominent entrance as a long wall juts out of plane to gather the students. The transition from the classroom to the chapel happens as students make their way through the plaza and up three short steps to the entrance. The public entrance is at the North West corner directed towards the parking lot. A small raised exterior foyer leading to a pair of large doors marks this entrance for ceremonial occasions.

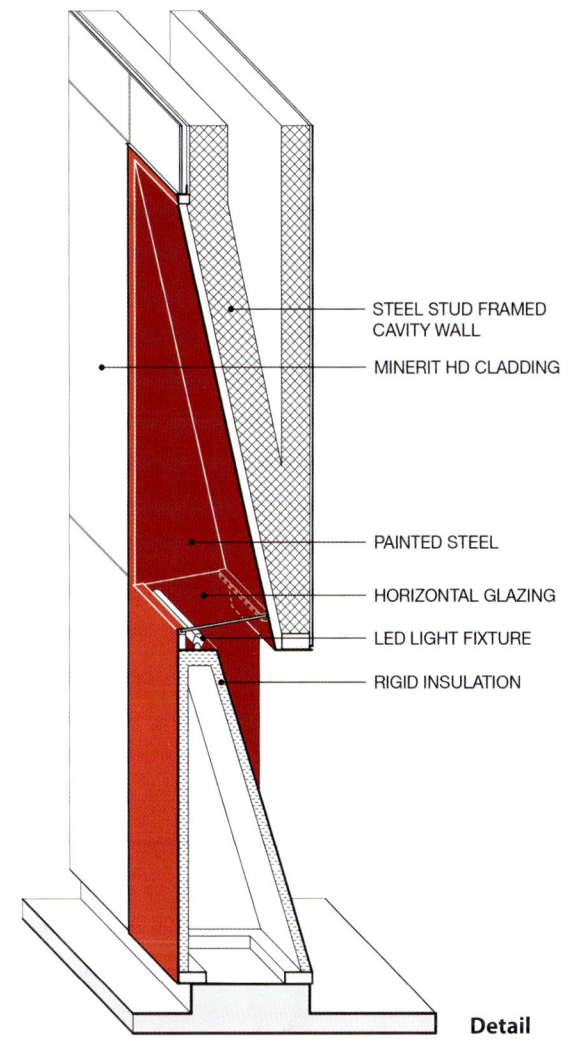

- STEEL STUD FRAMED CAVITY WALL
- MINERIT HD CLADDING
- PAINTED STEEL
- HORIZONTAL GLAZING
- LED LIGHT FIXTURE
- RIGID INSULATION

Detail

058

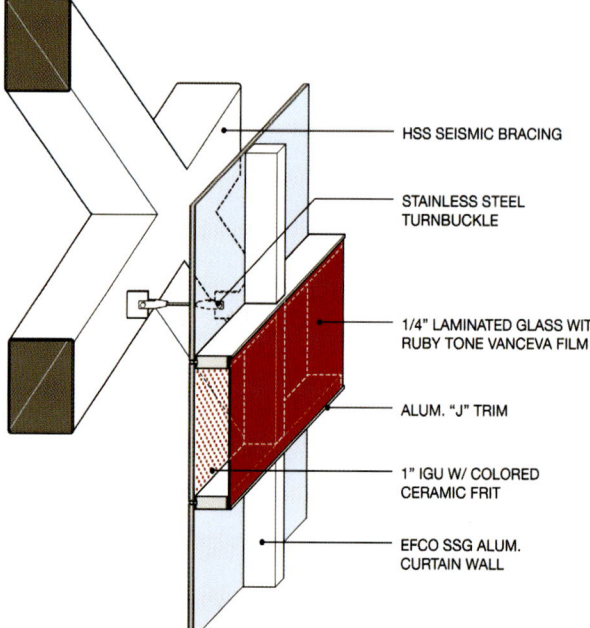

- HSS SEISMIC BRACING
- STAINLESS STEEL TURNBUCKLE
- 1/4" LAMINATED GLASS WITH RUBY TONE VANCEVA FILM
- ALUM. "J" TRIM
- 1" IGU W/ COLORED CERAMIC FRIT
- EFCO SSG ALUM. CURTAIN WALL

On the South side of the chapel a collage of transparent, translucent, and colored glass offers a glimpse of the vestibule to passers-by. Facing the campus, this is the most transparent face of the building. The east /west walls, clad in large cement board panels; act as anchors to the folded roof above. Above, a slim tower pierces the roof, to be surmounted by a simple, unaffected Cross, which will be visible from both the Campus and the surrounding area.

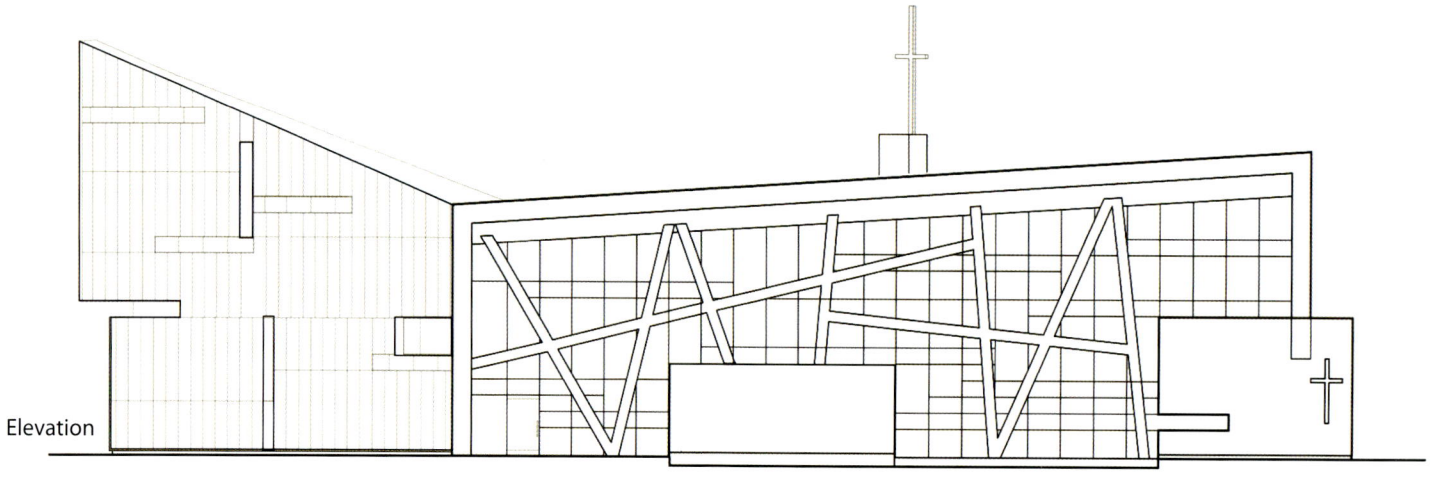

Elevation

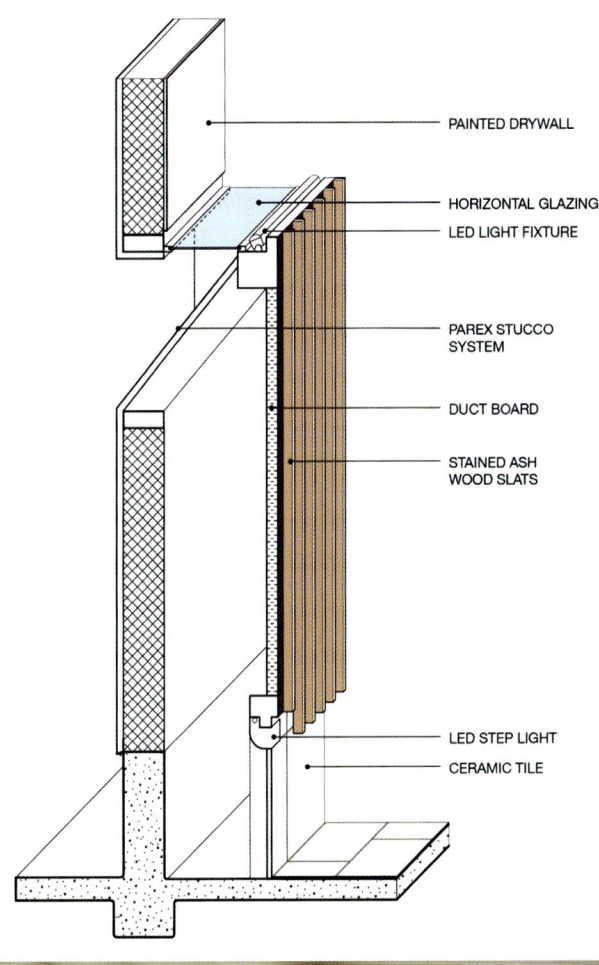

- PAINTED DRYWALL
- HORIZONTAL GLAZING
- LED LIGHT FIXTURE
- PAREX STUCCO SYSTEM
- DUCT BOARD
- STAINED ASH WOOD SLATS
- LED STEP LIGHT
- CERAMIC TILE

060

Housing in Paris

Architects: Projectiles
Location: Avenue de France, 75013 Paris, France
Area: 1250.0 sqm
Project Year: 2015
Photographs: Vincent Fillon

The project consists of two volumes, one on top of the other. An oblique section of the façade gives the project a kinetic aspect when approached from Avenue Secrétan.

The unified outer surface expresses our wish to give the whole building an elegant and sober character; it is also a way of limiting the multiplication of the diverse architectural elements.

The wall covering, using perforated aluminium boxes, contains all the light-modulation shutters of the residence and acts as a privacy filter, while decorating the outside areas such as the loggias of the family apartments or the technical terrace.

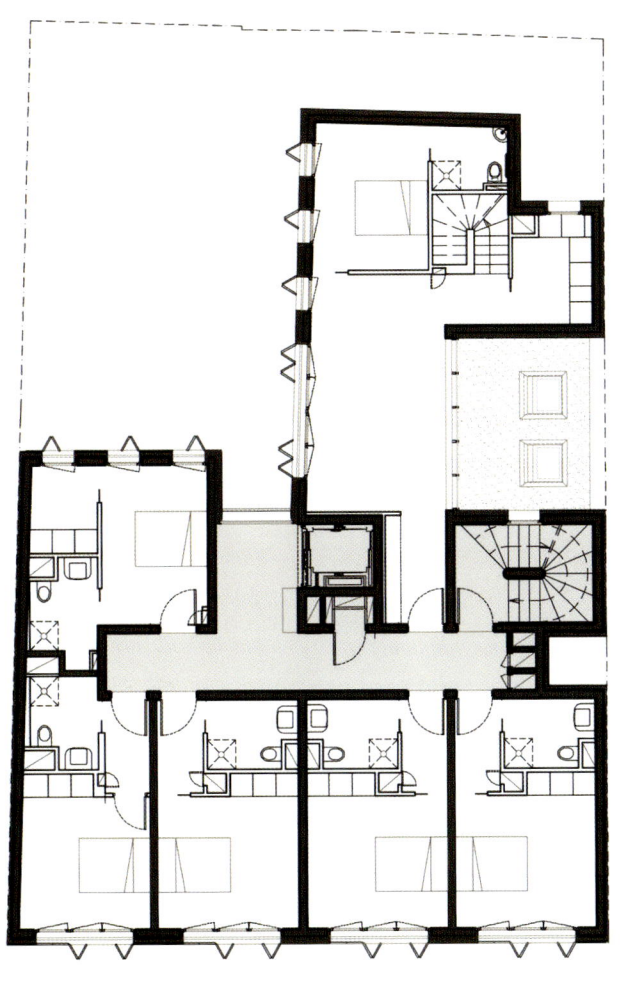

Floor Plan

0 1   5   10

062

Hotel Mercure in Bucharest

Architects: Arhi Group
Location: Bucharest, Romania
Architects in Charge: George Mihalache, Bogdan Stoica
Area: 6500.0 sqm   **Project Year:** 2015
Photographs: Cosmin Dragomir

Structural Design: Zagaican Birou de Structuri
MEP Design: MC General Construct
General Contractor: Conarg (local Romanian company)
Facades Contractor: Moldivars (local Romanian company)
MEP Contractor: Modulex Group (local Romanian company)

Hotel Mercure Bucharest City Center is a building with a story behind. The story appeared as a design consequence and has been further used as a feature for the hotel, which is a story-telling type. As it is situated in a historic and eclectic neighborhood, but very central, we've had several attempts to find means of stylistic expression, until we decided that the future building must contribute synergistically to the character of the area, which led to a contextual approach. The hotel facades have taken and reinterpreted architectural and symbolic elements from the immediate vicinity, so as to emit a cultural message, both urban and stylistically consistent.

The houses on this particular street were an inspiration for the arched windows, which we alternated with rectangular ones, in a playful manner; also, we reiterated one of the Romanian Athenaeum's main architectural features, the portico, which is detached from the main volume and ceded to public space. Last but not least, complementary to these archetypal items, on the main facade we have inserted a graphical collage of cultural symbols that takes the story further.

Elevation 1

Elevation 2

The technical process we used to accomplish the modern "fresco" is a very actual architectural solution: we used CNC perforated panels, placed at a certain distance from the facade support, creating the look and feel of depth by alternating light and shade. The hotel consists in some 6500 sqm of GBA, 2 basements and 7 floors above ground, having 114 rooms.

Ground Floor Plan

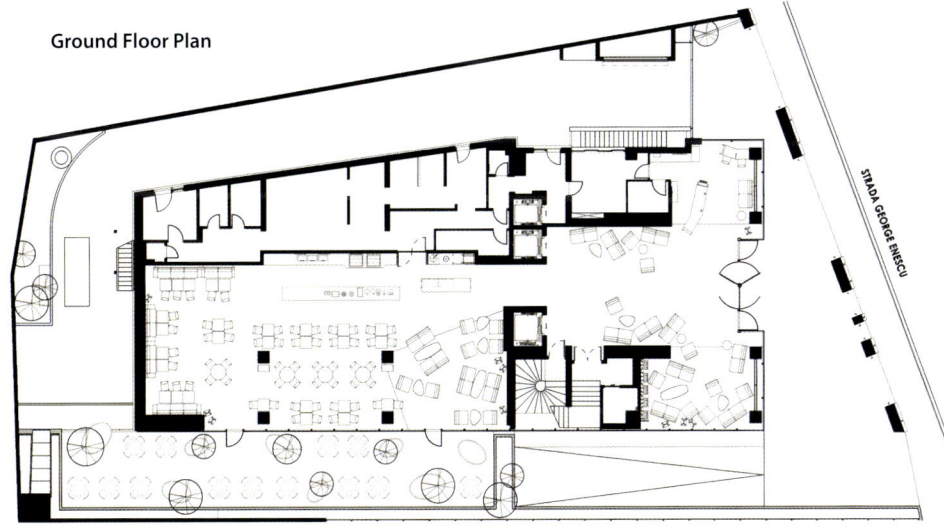

STRADA GEORGE ENESCU

Upon a crop of the entablement of the Romanian Athenaeum floats a pair of muses "descending" from the ceiling of the music salon of the museum George Enescu - double tribute to two great local symbols: Romanian Athenaeum (50m away from the hotel) and the great composer George Enescu, which gives the name of the street on which the hotel is located.

The technical process we used to accomplish the modern "fresco" is a very actual architectural solution: we used CNC perforated panels, placed at a certain distance from the facade support, creating the look and feel of depth by alternating light and shade. The hotel consists in some 6500 sqm of GBA, 2 basements and 7 floors above ground, having 114 rooms.

Chuon Chuon Kim Kindergarten

Architects: KIENTRUC O
Location: District 1, Ho Chi Minh, Vietnam
Architect in Charge: Đàm Vũ
Area: 486.0 sqm   **Project Year:** 2015
Photographs: Hiroyuki Oki

Design Team: An Ni Lê, Triết Lê, Tiên Đặng, Nhung Hồ, Dân Hồ

This project is a conversion of an existing town house into a private kindergarten in Tân Định ward, one of District 1 subdivision, Sài Gòn, Vietnam. Multiple functional needs within a space demand a flexible spatial adaptation that allow the room to shift from a single private space to a larger public or event space, or to proportionally expand and contract in size while complying with the house's existing structural elements.

The architects approach this project with an understanding children naturally feel more comfortable in spaces that are relatively related to their size, of which offer a sense of safety and freedom to explore their surroundings.

From a larger point of view, a cluster of small spaces stimulates and encourages the children to go out and discover what is beyond their own personal bubble. The flexible spatial organization is a direct respond to programmatic requirements and the educational experience the architects want to offer.

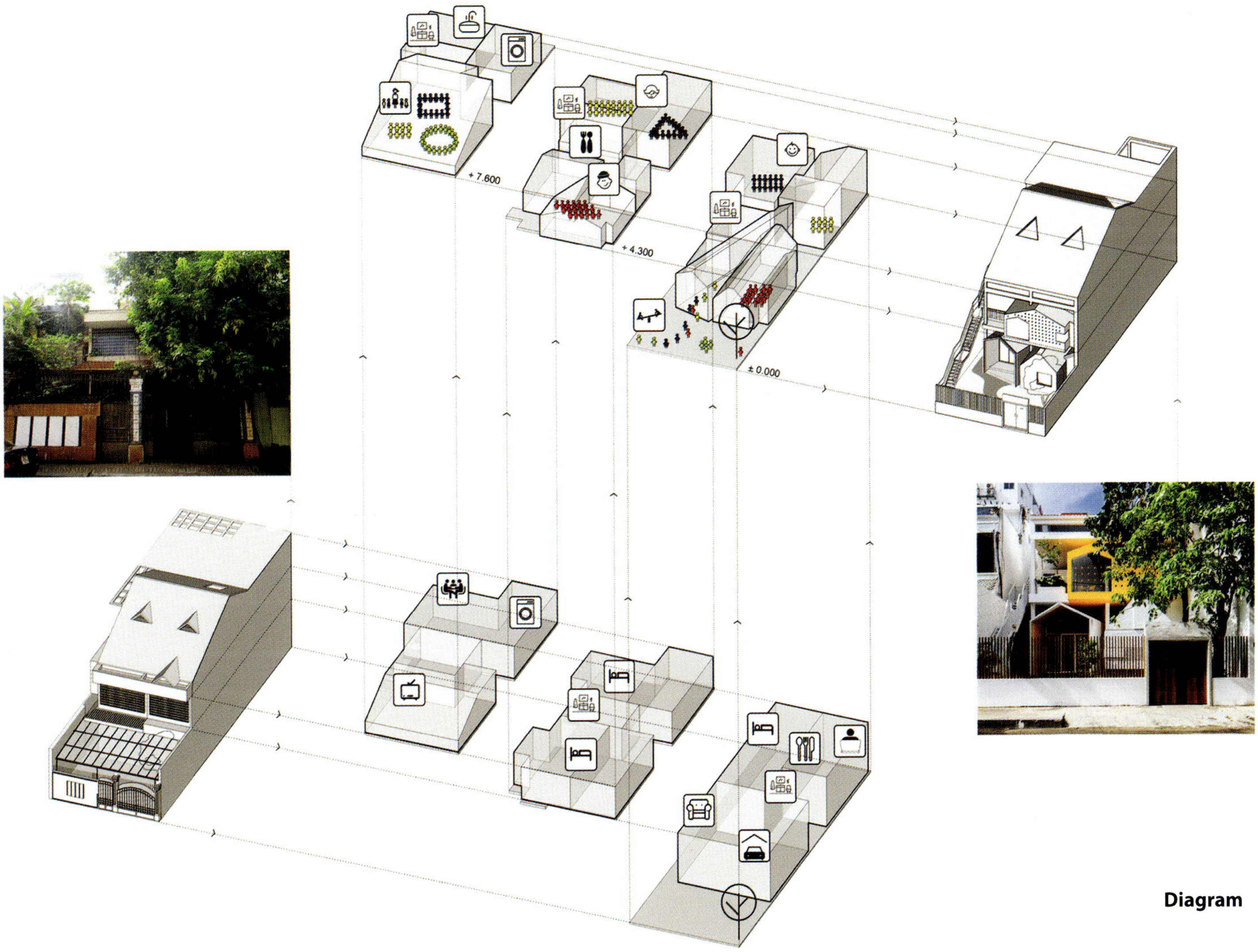

Diagram

A common practice in developing country, whose society progresses at a fast pace, is to deliberately recondition an old existing building and give it new functional or aesthetic purposes.

This is normally done accompanying a more complex spatial demands, limited resources, and time. Despite the constrains, the team have transformed an old, and monolithic space to a bright and exuberant vitality kindergarten, a place that foster the new generation of Vietnamese people.

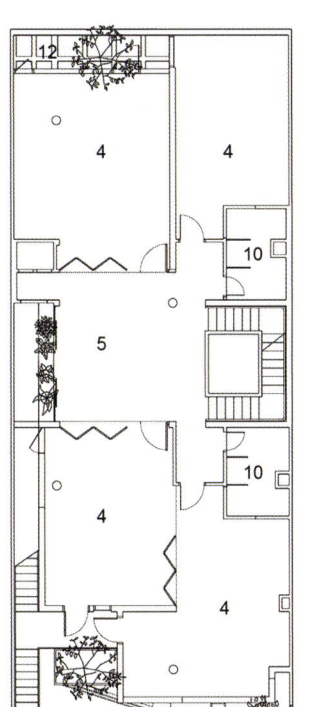

1F PLAN

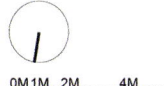

1. RECEPTION
2. STAFF
3. RECOVERY ROOM
4. CLASSROOM
5. FREE SPACE
6. ART CLASSROOM
7. KITCHEN
8. PLAY GROUND
9. TERRACE
10. LAVATORIES
11. LAUNDRY
12. BACK YARD

Perspective

The Forum

Architects: Studio804
Location: Lawrence, KS, USA
Area: 3000.0 ft2   **Project Year:** 2015
Photographs: James Ewing Photography

Collaborators: Nathan Brown, Renee Brune, Krista Cummins, AJ Dolph, Nicolas Elster, Jordan Goss, Ken Grothman, Christine Harwood, Ian Heath, Sara Lichti, Michael McKay, Josh Ostermann, Tim Ostrander, Ben Peek, Alyssa Sandroff, Aaron Sirna, David Versteeg, Jonathan Wilde

The Forum, located in the Architecture department of University of Kansas, was designed and built by students to meet the United States Green Building Council's Leadership in Energy and Environmental Design standard, LEED v2009 NC Platinum. This is the most stringent new-construction certification for sustainable-design that was in use by the USGBC during the building construction. It marked Studio804's seventh consecutive LEED Platinum building with several other outstanding architectural features. It includes a unique positive displacement ventilation system. On mild days during the year, the natural ventilation mode shuts down both the primary and outdoor air system, and cross ventilation is used to draw in fresh air through the space.

The perimeter skin is made up of two separate walls of insulated glass, three-and-a-half feet apart. The space between them provides room for the cedar louvers. Motorized dampers are located at the top and the bottom of the outer layer which permit cool air to enter in the summer when open. The louvers are controlled by a rooftop weather station and are programmed to track the sun. The living wall of ferns and begonias provides additional acoustic control for the lecture space by absorbing and dissipating sound and is watered from the cistern.

The lush vegetation helps to improve indoor air quality by naturally filtering toxins in the air. Besides these, setting up the LED lights, PV panels, water reclamation and usage of recycling materials are the other great aspects of this building. The construction had begun with the demolition of an outdoor workshop on the south side of the Architecture building. The existing shed housing the electric services had to remain in continual operation. As a result, the addition was carefully built above the existing mechanical room, and this became the site for the new glass-enclosed 121-seat auditorium and a 'Jury room' with a big common space at the entry.

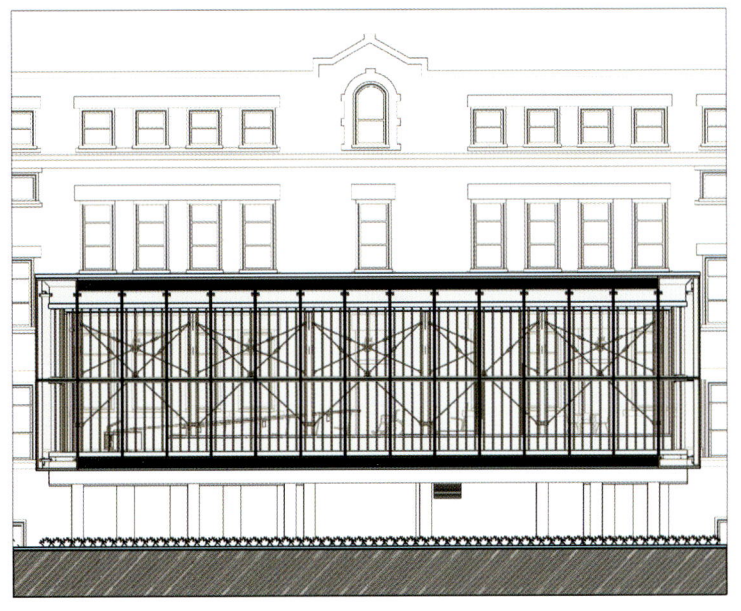
SOUTH ELEVATION

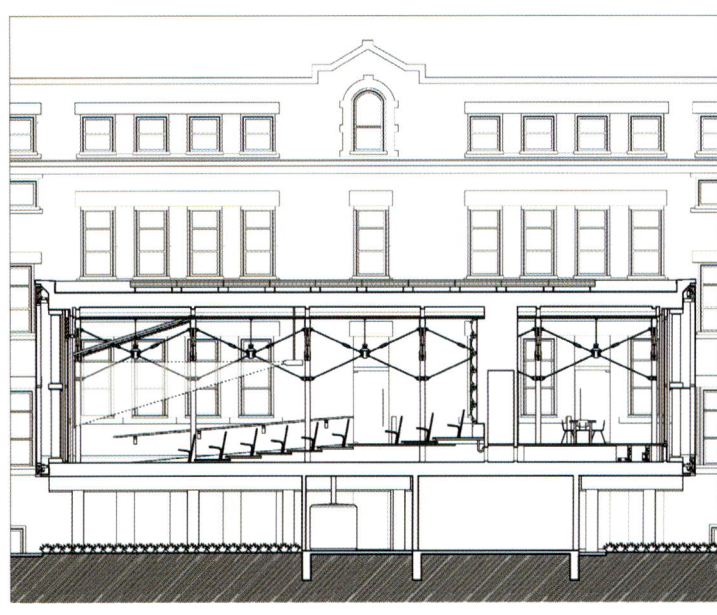
LONG SECTION

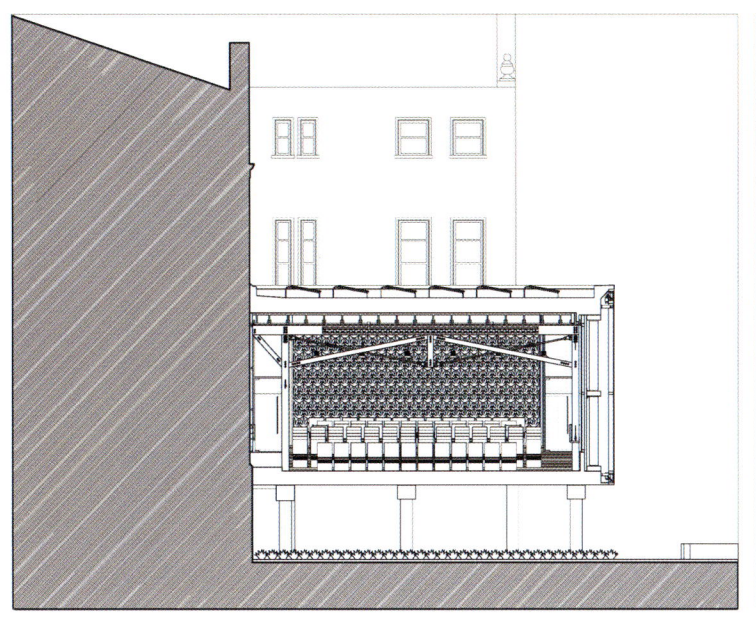

CROSS SECTION

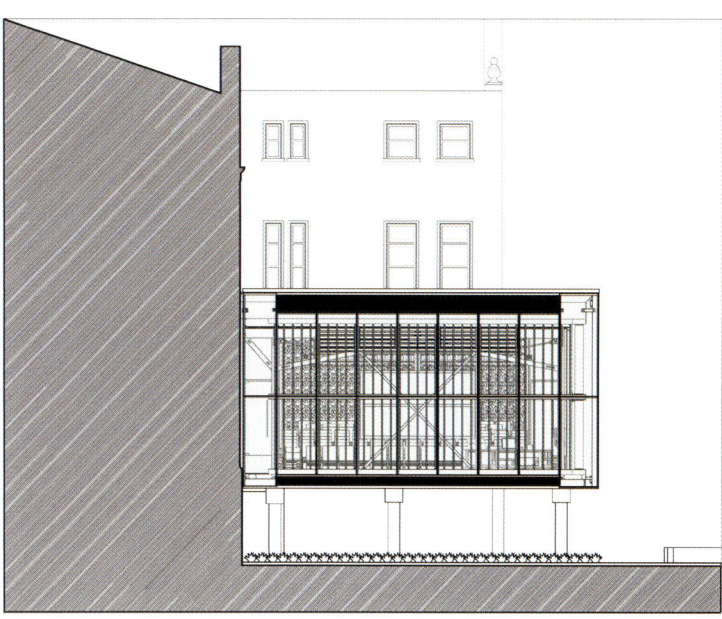

WEST ELEVATION

078

Nasu Tepee

Architects: Hiroshi Nakamura & NAP
Location: Tochigi Prefecture, Japan
Area: 156.0 sqm
Photographs: Koji Fujii - Nacasa & Partners Inc, Courtesy of Hiroshi Nakamura & NAP

Tochigi Prefecture, a well-known summer resort. Passing through fields and woodlands, the site lies along the forest path in a grove of mixed trees. Our client is a married couple that enjoys organic farming on the weekends, and their wish was to reserve as much of the environment as possible and to live in the surrounding woods. We avoided large-scale construction and a majority of felling to build the rooms on the few remaining flat surfaces of the sloping ground, as if sewing them together.

Since the site is located in the midst of dense woods, we came up with the idea of having a high ceiling to let direct sunlight in the house. However, this would result in the space becoming too large, cost a lot to run the air-conditioning and the exterior walls would bump into the branches. Therefore, we eliminated unnecessary space. First, we cut down the upper space diagonally to make the ceiling lower based on the way people move. In addition, the new form matches the tree branches that spread out radially. This resulted in a tent-shaped house with only one third of the volumes. Although the highest ceiling is 8m, the average ceiling height dates a standard 2,6m. As a matter of fact, people cannot stand upright close to the walls, so we simply turned the spaces into sleeping and sitting areas. The ceiling descends like a tent and enables the creation of a warm living space that mingles with the trees. You will feel the warmhearted embrace of the house around you.

LONG SECTION

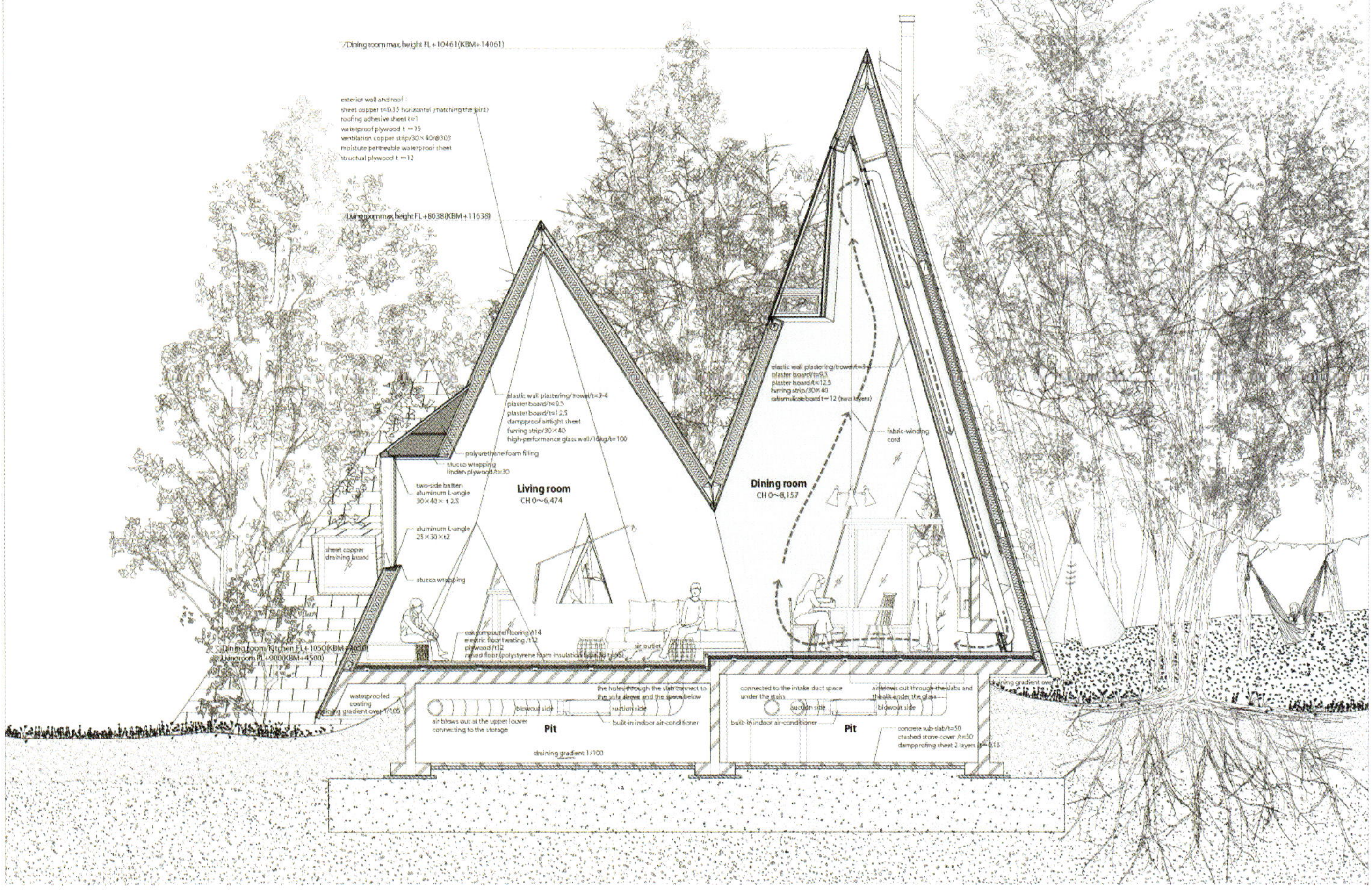

It is similar to primitive spaces seen in the houses of the Jomon People (Ancient Japanese) and Native Americans. The structure of the house initiated a lifestyle with close interaction, because the family sat along the low wall facing each other. A fire, a light or a table was set in the middle to initiate conversation as the family gathered around the center. The architecture has had an influence on people's habits and it strengthened the connection and bond of the family. At night, the moonlight becomes the only natural illumination and the dark forest falls into silence. At times, the presence of wild animals can be heard ever so faintly. In the darkness of the night, you will find a house filled with warm and gentle looks. The "pointy-hat" building cantilevers slightly above ground in order to prevent insects, humidity and fallen snow entering the house.

The windows are all double glazed in order to ensure that the tall spaces are airtight and well insulated. We installed the fireplace and the air-conditioning underfloor by making use of the floor heating and the pit. During summer, warm air gathers at the top and escapes through the top light side window. During winter, warm air at the upper part will be drawn in and blown out at floor level. This way, we are able to create a very comfortable air environment.

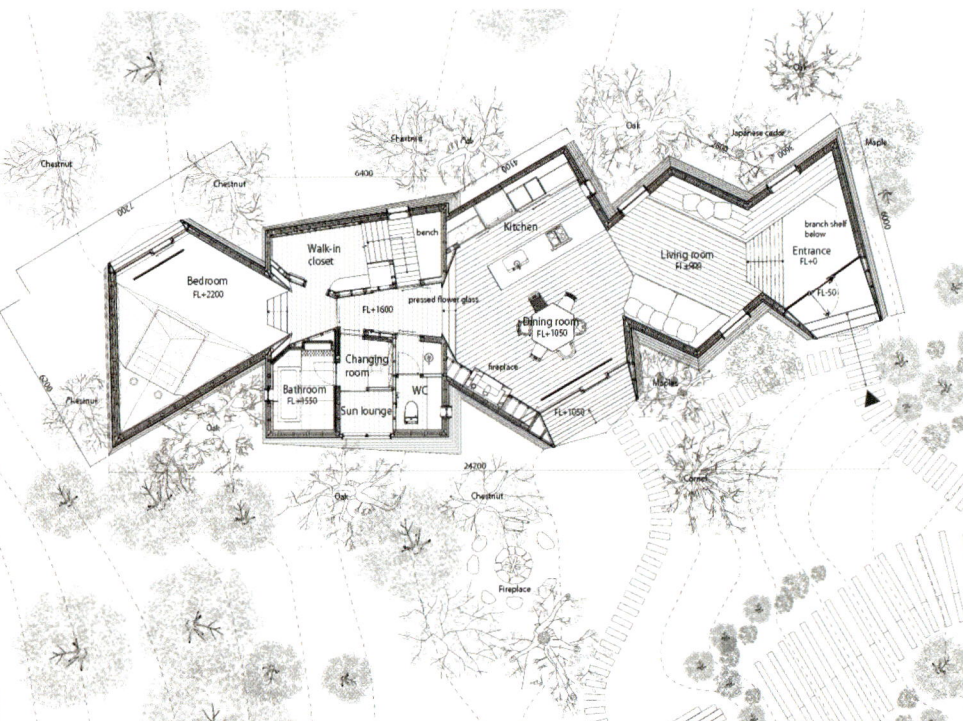

Floor Plan

Akebi, viola, anemone, geranium, larkspur – the wild flowers found in the pressed flower glass all came from the surrounding woods. Our idea was to find a new way to reflect the blessings of nature, not just in the context of samples or picture books. We manufactured the glass by sandwiching the pressed flowers in resin films between two 4mm thick glasses and firing them in vacuum. We put films for ultraviolet protection in the glass to protect the flowers from decolorization.

082

Concrete House in Caviano

Architects: Wespi de Meuron Romeo architects
Location: 6578 Gambarogno, Switzerland
Area: 128.0 sqm   **Project Year:** 2015
Photographs: Hannes Henz
Commitment: Fam. Jérôme, Paola de Meuron

Collaborators: Giorgi & Partners, IFEC Ingegneria SA
Master Builder: Canonica SA
Windows: Morotti GmbH
Plasterer: Paolucci SA
Joiner: Romeo Buss GmbH

The house, designed as a residence for a family of three persons, was built in the immediate proximity of the architecture office in Caviano on the Lake Maggiore. In terms of adequate architectural densification new living space should be created on a remaining area of just 128 m2, on the same plot as the architecture office was built in 1981, without damaging the existing qualities. On the contrary, an enrichment of the outer spatial situation should be generated with reasonable densification in context with the existing building. The building laws determined the outer form of the building, what often happens when leftover plots are developed. The minimal distance to the road, the minimal distance to the forest, the minimal building distance to the architecture office as well as the right to build on the limit to the southwest neighbour, create an irregular pentagonal form of totally 79 m² surface. A clear rectangle of 48 m² surface, which is the isolated interior, was integrated in this irregular form.

The polygonal exterior shape and the steep topography of the site let the building appear as an archaic stone block in middle of the forest, this is reinforced by the rough washed concrete surfaces becoming darker by the weathering.

Floor Plan

Roof Floor Plan

To the mountain-sided street the construction presents itself as a closed, simple one-storey volume. The only opening towards the street is the raw steel gate leading to the entrance court. A 3 m wide forecourt with a natural stone pavement and two palms connects the house to the street and upgrade it spatially. To the valley-side, the house appears as a narrow 3-storey tower. The house is organised on three floors: the top floor on the street level accommodates the entrance, the main living area and dining with the open kitchen, on two sides it's completely closed and on the other two sides it's completely vitrified towards the courtyards.

Both courtyards, each with a wisteria, let the living room becomes a "garden" room and let the inhabitants experience in an unusual intense way the varying atmospheres of the weather and the light. A skylight above the staircase allows light to penetrate into the lower floor, which accommodates two bedrooms, each with its own outdoor loggia, the bathroom and the stairs to the cellar, where is the technique and a workspace.

House in Tsudanuma

Architects: fuse-atelier
Location: Narashino, Chiba Prefecture, Japan
Area: 63.0 sqm   **Project Year:** 2014
Photographs: Shigeru Fuse
Architect in Charge: Shigeru Fuse
Design Team: fuse-atelier, Musashino Art University / fuse-studio
Structure: Reinforced Concrete
Structural Engineers: Ysutaka Konishi
Main Contractor: Three F

The house is for a married couple in their forties, located in the city of Narashino. It is situated on the edge of a residential neighborhood and a commercial district that stretches form the station, with large-scale stores nearby. The house faces an eight meter wide street classified as an urban roadway, with busy traffic due to its proximity to the train station. The client wished for a concrete house, where all spaces are stringed together in sequence, while also maintaining the privacy and freedom of a spacious living space.

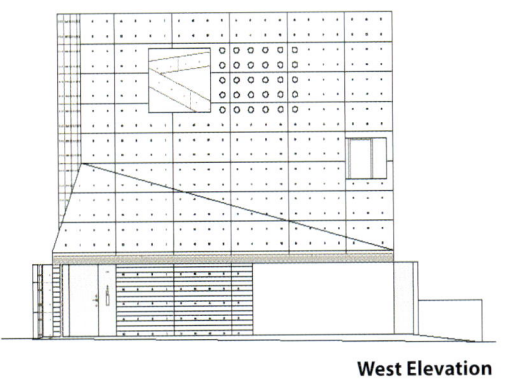
West Elevation

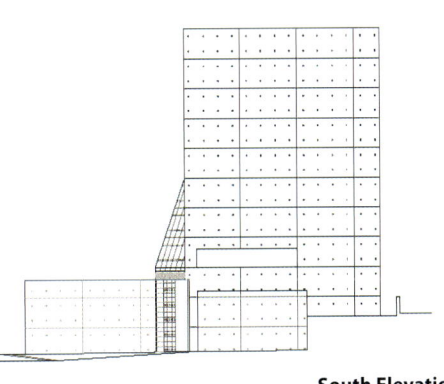
South Elevation

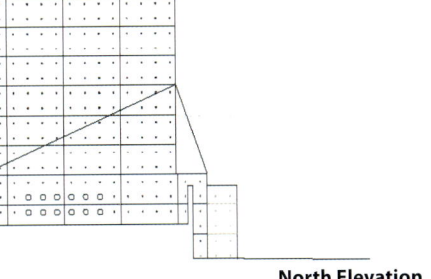
North Elevation

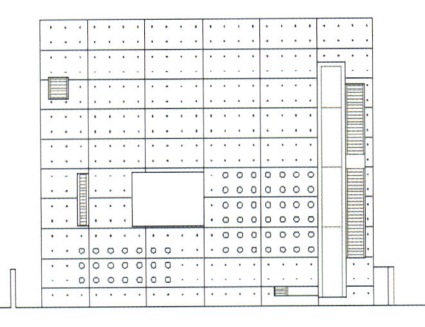
East Elevation

Since four meters of the adjacent road is specified as part of urban zoning, we built the structure in the back part of the site. The resulting open space in the front of the structure functions as a buffer between the street and the house. Concerned with the privacy and noise from the outside, we incorporated an outdoors space into the structure. Upper parts of the building structure had to be set back due to various legal limitations. As a result, the exterior of the structure was tapered from the second to third floors.

The various spaces of this house are interconnected by producing as many lines of sight and flow as possible. The vertically inserted outdoors space with a passage that leads from the second floor terrace to the rooftop effectively breaks up the interior space. Combined with the flow planning of the interior, this also amplifies the building's overall ease of navigation.

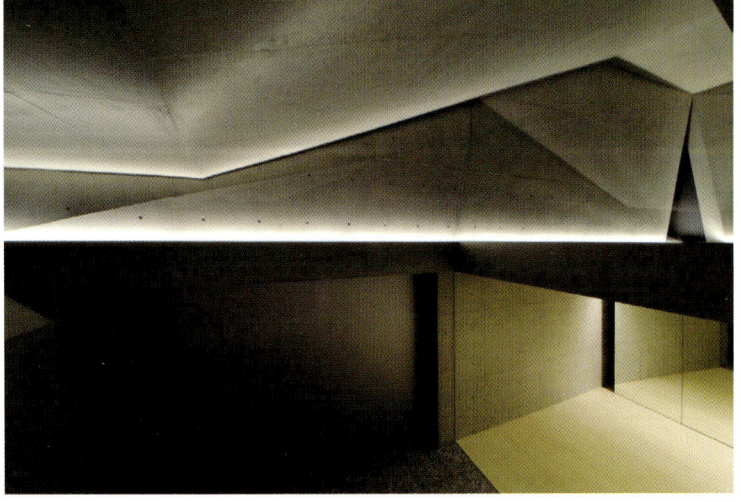

In addition to the flow planning on the exterior and interior, we layered a network of sight lines to inspire diverse spatial relationships. An accumulation of associated parts create gaps both horizontally and vertically, leading the eye through surprising escapes that result in a complex layer of various sequences. By fully embracing the minimalist details of concrete, glass, metal and rock, the space is intensified with distinct sharpness, where the skeletons of the spatial structure becomes visibly apparent. The house becomes a dynamically linked, three-dimensional chain of spaces that inspires various scenarios and distances.

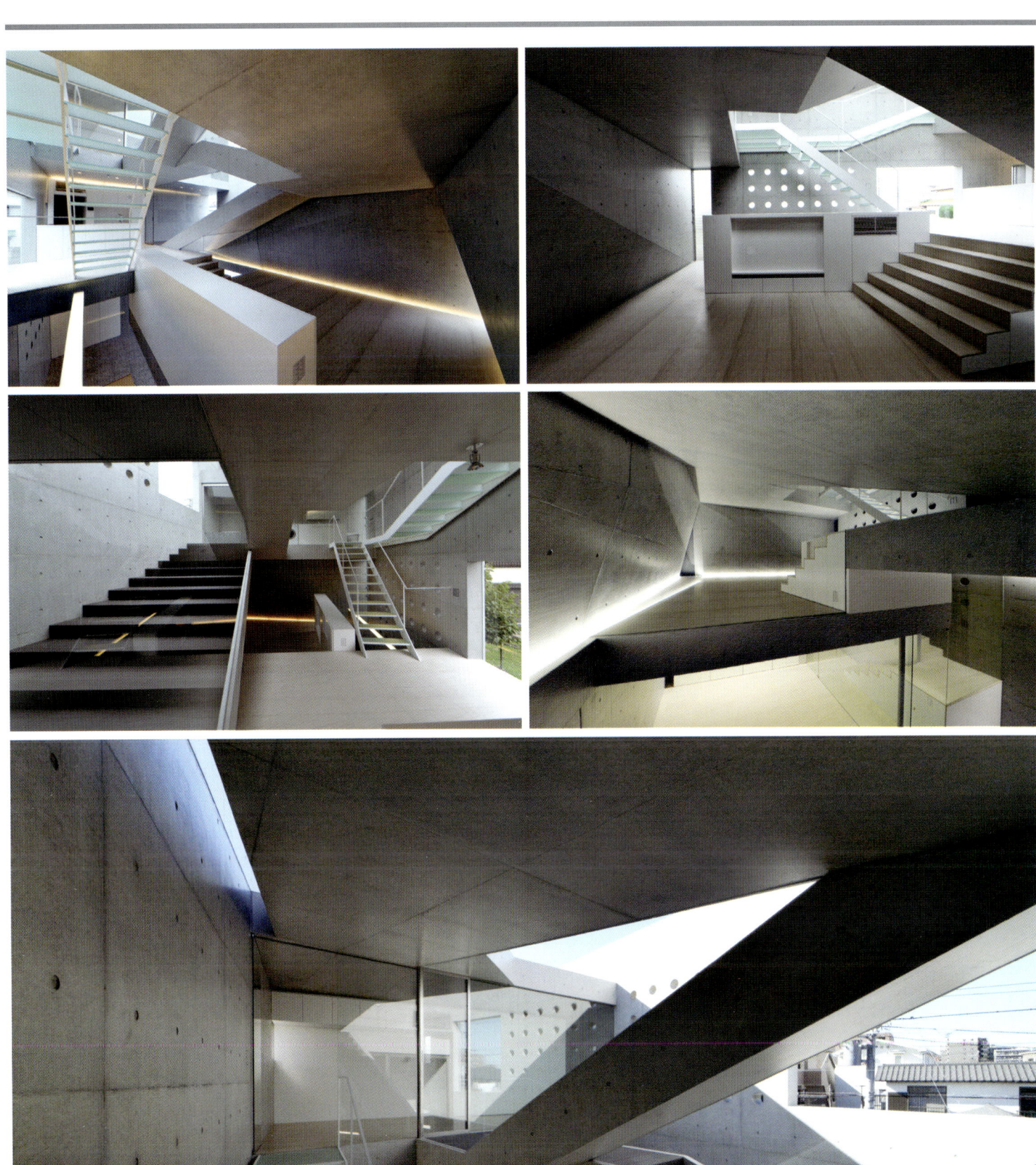

Golf House

Architects: Luciano Kruk Arquitectos
Location: La Costa Partido, Buenos Aires Province, Argentina
Area: 274.0 m2   **Project Year:** 2015
Photography: Daniela Mac Adden
Project Manager: Luciano Kruk
Project Coordination: Ekaterina Künzel
Site Management: Pablo Magdalena
Collaborators: Josefina Perez Silva, Andrés Conde Blanco, Federico Eichenberg, Dan Saragusti, Isabelle Ducrest

Golf House is geographically located at the center of Costa Esmeralda, a neighborhood 13 km north of the seaside resort of Pinamar. Topographically lying on highest area of the neighborhood, both the front and the back of the land plot adjoin a golf court that, along with the —mainly wild— native vegetation, compose its immediate surroundings. Originally, the plot of land was circumscribed by a virgin dune, the slope of which grew in height away from the street, and that the Studio proposed to preserve.

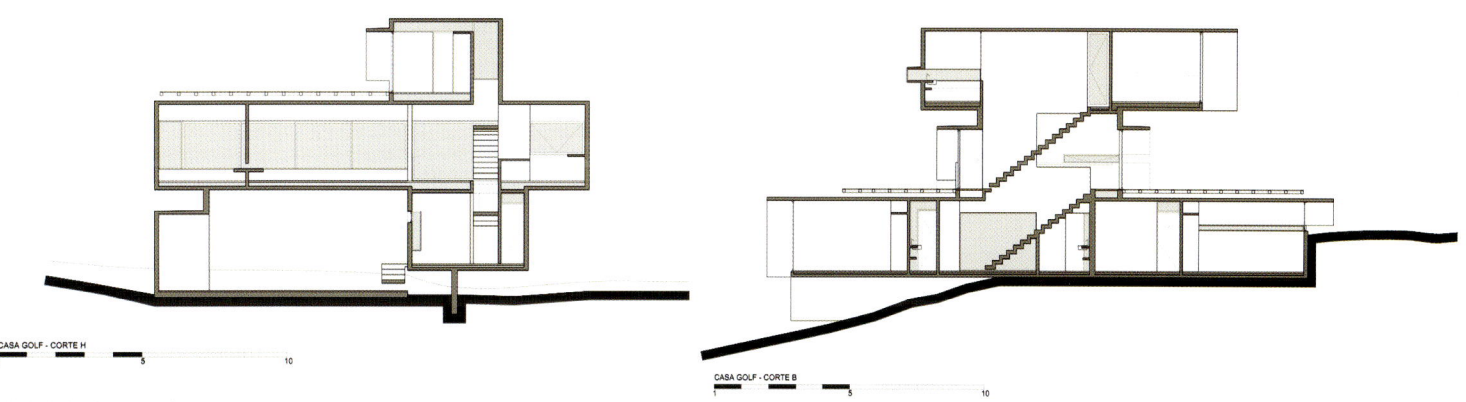

The Studio proposed a house entirely materialized in exposed concrete, whose noble esthetic expression allows a respectful dialogue with its surroundings. The architectural strategy aimed at organizing the different functional requirements distinctly grouped in three pure volumes oriented independently of one another and set at different levels. With its back half buried under the dune, the lower volume lodges the entrance lobby and the secondary bedrooms. Along with a cube standing opposite —serving as a warehouse—, it supports the prism destined to house the dynamics of the family's activities. Containing the master bedroom, the third volume stands at maximum height, thus satisfying the client's requirement of privacy. Regarding the general project, it aimed at exploiting the views, and at the same time avoiding a loss of the necessary preservation of the rooms' intimacy, which is a characteristic challenge of contemporary glazed architecture.

The volume housing the social areas is the most visually permeable one, not just due to the transparency of its skin, but also due to its being parallel to the street (contrary to the other two, that were set perpendicularly) and falling back into the lot. The social prism occupied the maximum building width and rose to the highest level of the dune at the back of the plot of land, thus gaining 180° views over the golf court, open views over the neighborhood, and an appropriate concealment from the street.

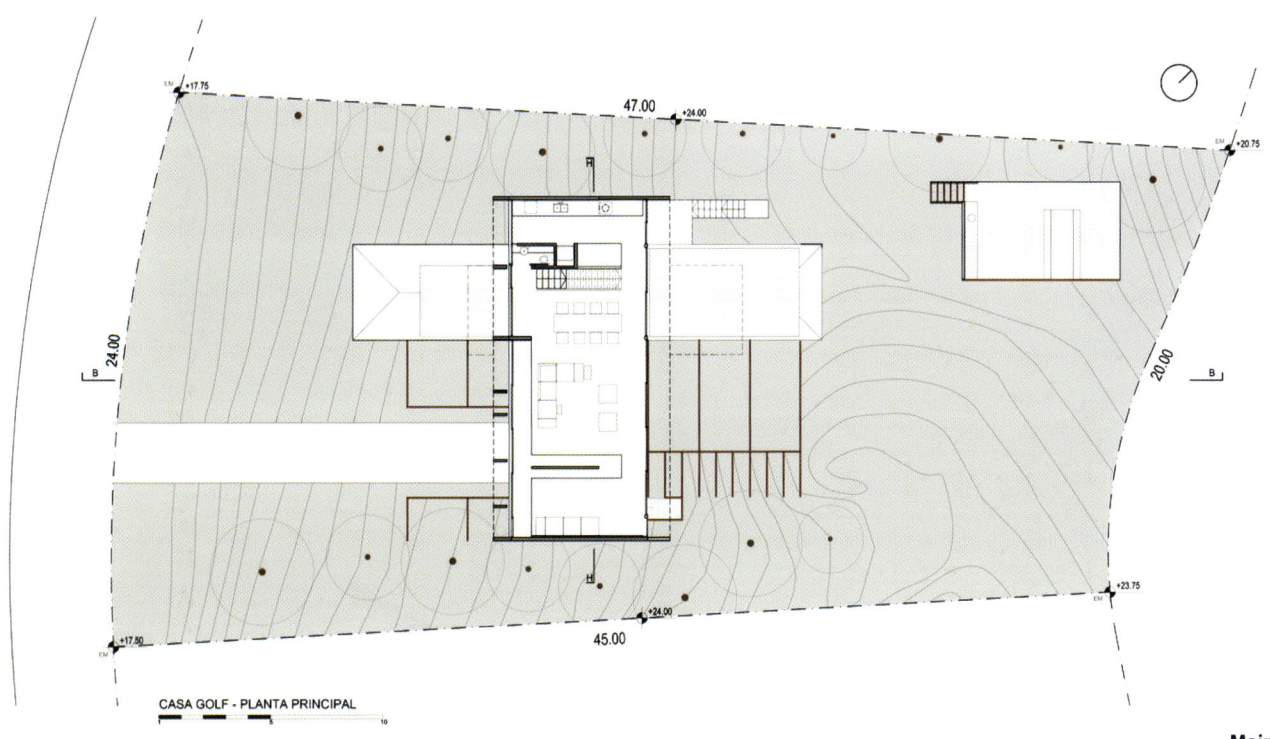

Main Plan

In order to better reinforce the volume's privacy, it was decided that maritime pines should be planted in the space lying between the building and the road and vertical sunshades be installed, that also reduced the sunlight incidence from the west inside. Likewise, the sunlight vertical incidence from the north was controlled by the horizontal eaves —cantilever slabs—, which monolithically joined to vertical partitions work in the same way as the brise-soleils at the front. Adequately independent from the rest, the third volume, like a cannon, captures views of the distant sea horizon. While the social area opens frontwards into a terrace-lookout, a bigger, more private expansion on the other side projects the living room and dining area outwards. Providing shelter from the sun and the rain, the master bedroom's volume hangs over most of this expansion. Every deck was built with quebracho wood and contributes to lower the sun incidence over the rooms they lie on through their being slightly detached from the concrete.

The facing walls less affected by the sun rays (oriented to the southwest) were thermally isolated with an interior coating made of kiri wood. This coating was also used on the partitions that support the backrests of the beds, so that the concrete's roughness is softened in the most private areas. Besides passive sun control provided by the house's architecture itself, and in order to allow Golf House to be lived in throughout the year as the commissioner requested, Split air conditioners and radiating floors were installed. The prisms' disposition aimed at structuring the house as a lookout-artifact composed of volumes set around an articulating axis: vertical circulation. The scope of their overlapping and the partially underground entrance lobby was to lower the height and moderate the visual impact of the whole building. The big spans and the overhangs called for by this volumetric distribution were only possible by means of the employed reinforced concrete's structural properties.

Fab-Union Space On The West Bund

Architects: Archi-Union Architects
Location: Xuhui, Shanghai, China
Architect in Charge: Philip F. Yuan
Area: 368.0 sqm **Project Year:** 2015
Photographs: Hao Chen, Su shengliang
Design Team: Alex Han, Xiangping Kong, Xuwei Wang
Structural Engineer: Zhun Zhang

The project which is supposed to be an impressive practice in the city is designed by Philip F. Yuan and Archi-Union Architects. Although it is micro in scale, FU Space is powerful enough to represent a new attitude to the value shift in architecture. It is located in the West Bund area in Shanghai, which is planned to be a future art and culture center of the city. Adjacent to ShangART Gallery, Shanghai Art fair Center and several other architect studios, the location of the site is terrific It is only 200 meter away from Huangpu river front and 2 blocks away from the historical quarter of Longhua Temple. Moreover, Long Museum, Shanghai Photography Museum and Yuz Museum are all within walking distance. It's undoubtedly among the art community which is taking significant experiments in Shanghai.

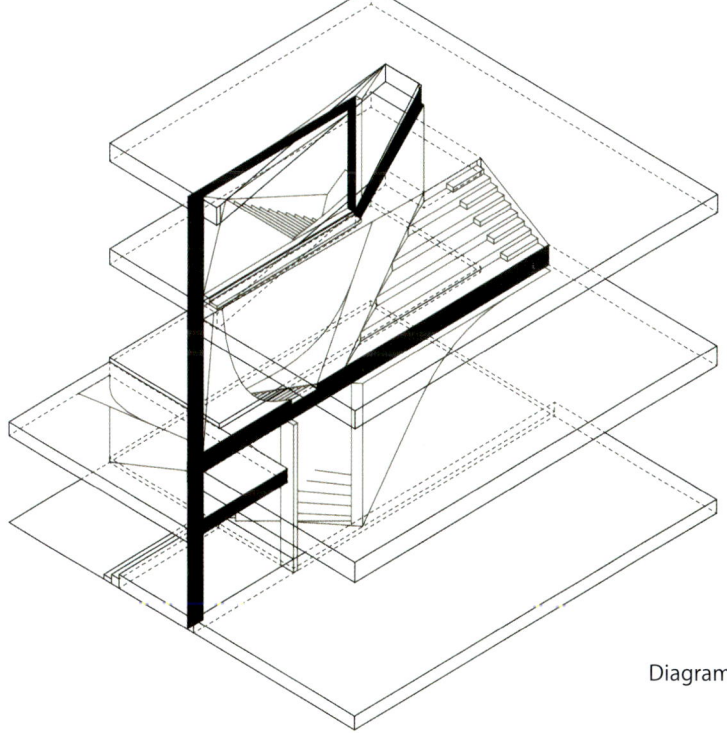

Diagram

FU Space will become a future non-profit contemporary art, architecture and culture communication center. It is aimed to be an exhibition and communication space. The site is very compact, located at a sharp turning corner. N Both the different circulations from 3 directions and a connection to the 2nd floor platform of Shanghai Art Fair Center have to be considered. Therefore, the primary concept is to set up a good soft joint for the whole community. The analysis leads to a form finding process throughout the geometry of the site. The inspiration of the material to reach this softness comes from the concrete of the platform, which originally cold and tough. If we use the mould system, we can actually implement the concrete to any soft surface. The program is specially set for exhibition. Five basic spaces including two 4.2M height space and three 2.8M space are all regular square ,which could be flexible for multi-functional purposes. .

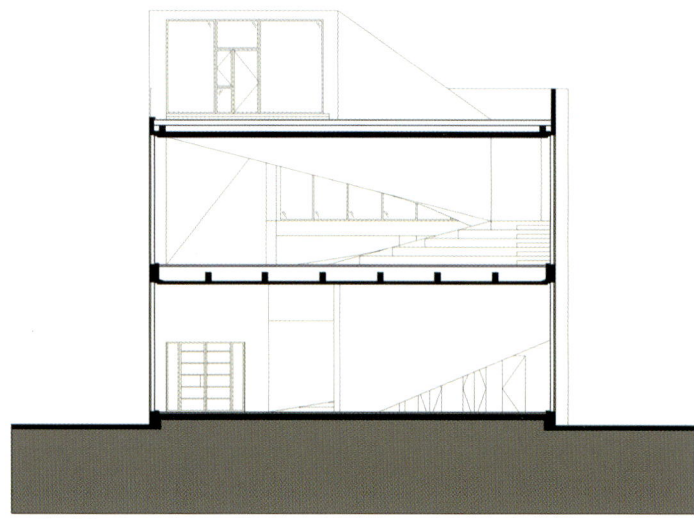
section 1-1

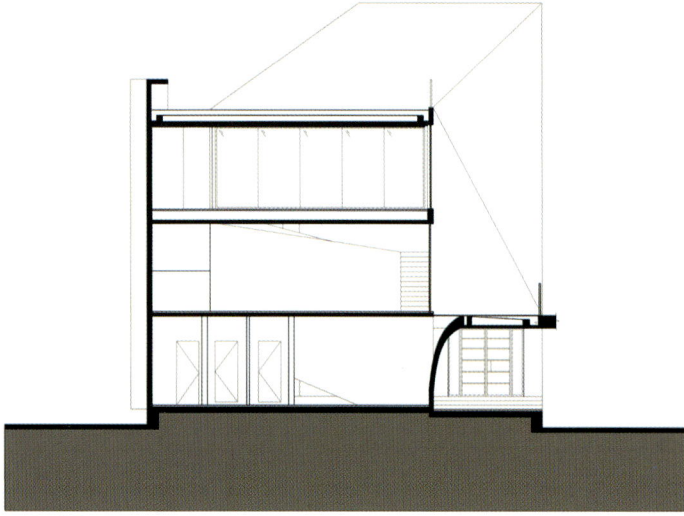
section 2-2

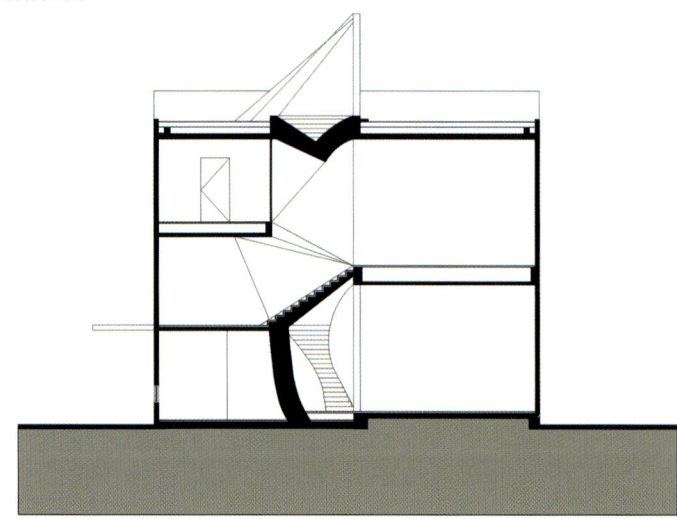
section 3-3

All spe cial space experience lies in the in-between circulation space. Interior public space followed by the exterior form finding process, is enhanced by an abstract thinking on creating a kind of experience climbing the rockery of Chinese garden. The key aspect of Chinese garden design is to make it big through small scale, which is extremely efficient in changing sceneries with varying viewpoints. The abstract process for this kind of experience is achieved by metry, which is intensified from different perspective.

The construction process was conducted in a very short time. Although all the unsatisfied marks and traces were recorded on the concrete wall, a real diversity strengthened a new sense of place in such a compact space. The softness of concrete, which makes the light flowing leisurely, touches the depth of heart of the visitors. The first exhibition of FU Space is especially dedicated to the chief architect, Philip F. Yuan exhibits his thinking, working models, and a topic: Presence of Absence, hopefully that could be an explanation for the concept of the building. The primary consideration in the conception of this project is to ensure the exhibition buildings on the side relatively complete. But the road space of 3 meters is built on the basis of the dynamic behavior of people, the air dynamics of the wind and the maximum volume of space continuity.

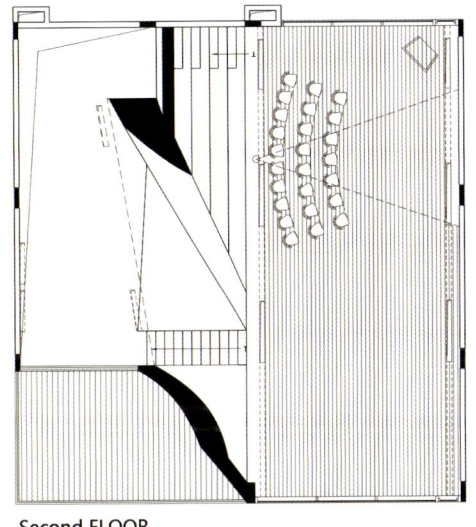

Second FLOOR
105.53 sq.m.

0 1   3   5m

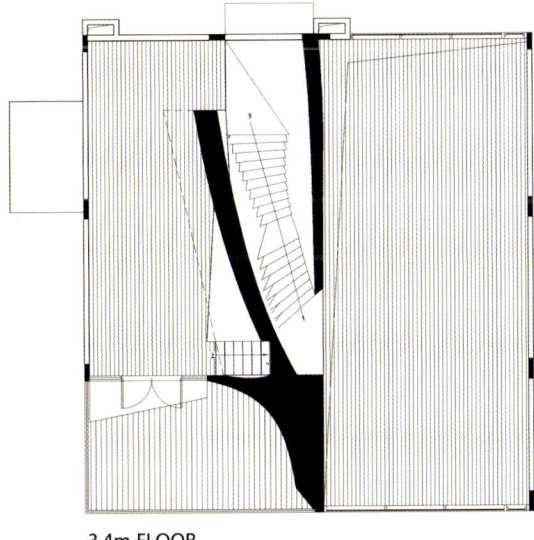

3.4m FLOOR
58.44 sq.m.

0 1   3   5m

Dynamic nonlinear spatial shape is built on the basis of structural performance optimization and spatial dynamics. The whole process use a variety of design methods of cutting stone, perspective geometry, and the Algorithm configuration.

Casa Premium Rama 2

Architects: PODesign
Location: Bang Khun Thian, Bangkok, Thailand
Area: 470.0 sqm **Project Year:** 2015
Photographs: Spaceshift Studio
Interior Designer: Is innovative studio
Landscape Designer: Quality Houses Public Company Limited
Structural: Chanapong Tongsen
Consultant: Quality Houses Public Company Limited
Cost Estimation: 10,000,000 baht

Casa Premium Rama2 is located on Kanchanapisek Road. A site is in triangular shape.
By using single plane wrapping around main space, we creates V-shape plan that engage spaces to the site. Building parameter is maximized to fit functions and visual. V-shape's lower wing becomes welcome gate and patio for swimming pool deck. Upper wing is inclined up to fitness room that can take view swimming pool below. Water fall sloping plane from lobby gradually transforms to center swimming pool.

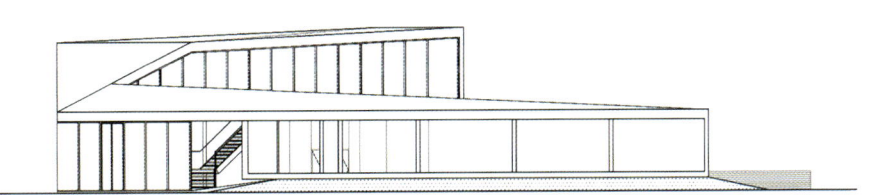

ELEVATION 1

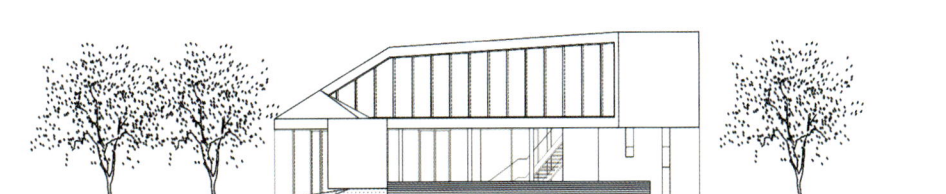

ELEVATION 2

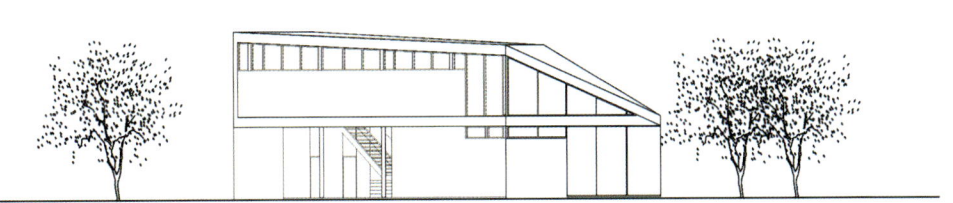

ELEVATION 3

Each wing of building casts shadow for pool area throughout the day. Entrance gate acts as an abstract frame leading one into the project in simple gesture.

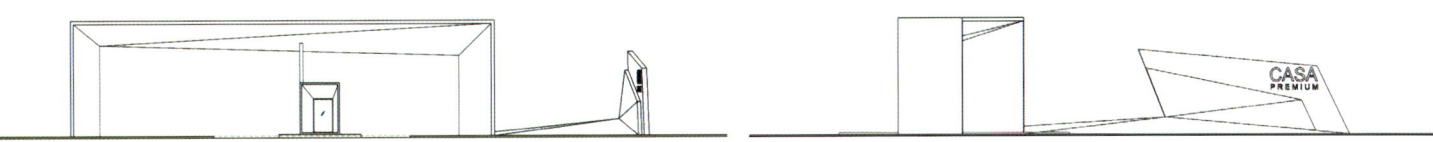

Gate Elevation 1 **Gate Elevation 2**

Filamentario Chapel

Architects: Architects Divece
Location: Tlajomulco de Zuniga, Jalisco, Mexico
Project Team: Paolino Di Vece Roux and Francisco Morales Dufour
Project Year: 2015
Photos: Jorge Silva

The architectural concept of Filamentario Chapel is to create a processional scheme, a perimeter route that favors the preparatory gathering for religious worship; the traditional, axial and finite scheme, which ends at the altar, is reversed to create a procession that considers the altar as the center of the composition.
The location of the chapel idealizes the duality of absence and presence; it is a building that is constructed but is also buried. It is like an archaeological find that forces us to make a downward journey, rewarded with the surprise of accessing from the back to keep the atrial cross as the axis of the project.

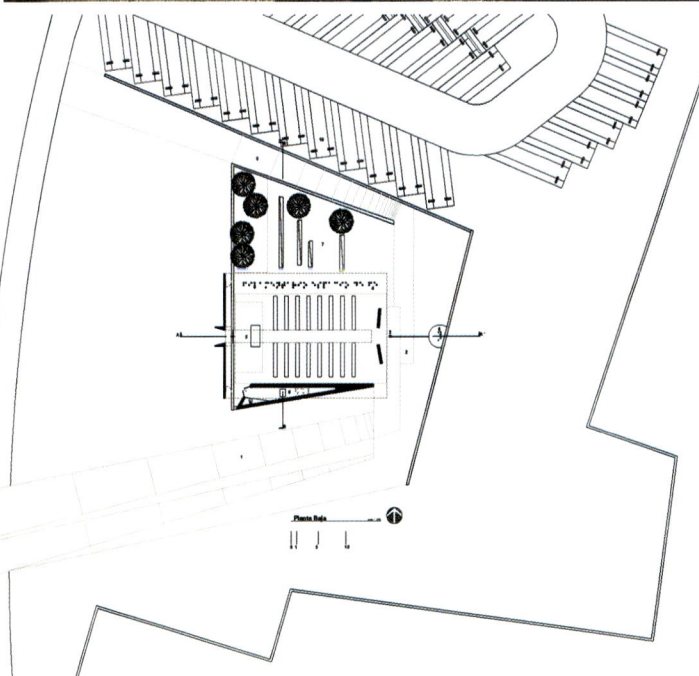

Ground Floor Plan

The altar, which is historically positioned at the end of the procession, is relocated to the geometric center of the project, at the foot of the atrial cross, to reinforce the mystical character of the volumetric complementarity and thus turn the cross into the summit and the origin of all dualities. The descending and perimeter routes around the cross are preparatory processions with their second contemplative station in the inner atrium, the entrance courtyard is splashed with the fountain of faith and in turn accompanies the baptismal pile as a preamble to the path inside.

Beyond the threshold of the chapel and with the cross at the front for a second time, the duality of the corporeal and ethereal is reinforced; The massiveness of the north wall contrasts with the lightness of the fragmented south wall; These fragile and spiky metal elements, almost ephemeral, not only contrast with the concrete walls due to their massiveness and lightness ratio but also due to the verticality of the opposed filaments, with the same geometry and proportion, as the texture of the concrete walls in the opposite direction. Moreover, these fragile filaments, which flood the interior with light, weave a close relationship between the outer edges and the covered space. The light enhances its presence and reveals the volume of absence, a manifestation that the corporeal and ethereal are one and the same.

At the height of the path, hidden inside the north wall, is the chapel of light; a brief space, a chapel inside the chapel, which marks the climax of the procession. The overwhelming darkness takes its full meaning when the few rays burst into the space to become highly contrasting elements and give supreme value to the mystical essence of interior light.

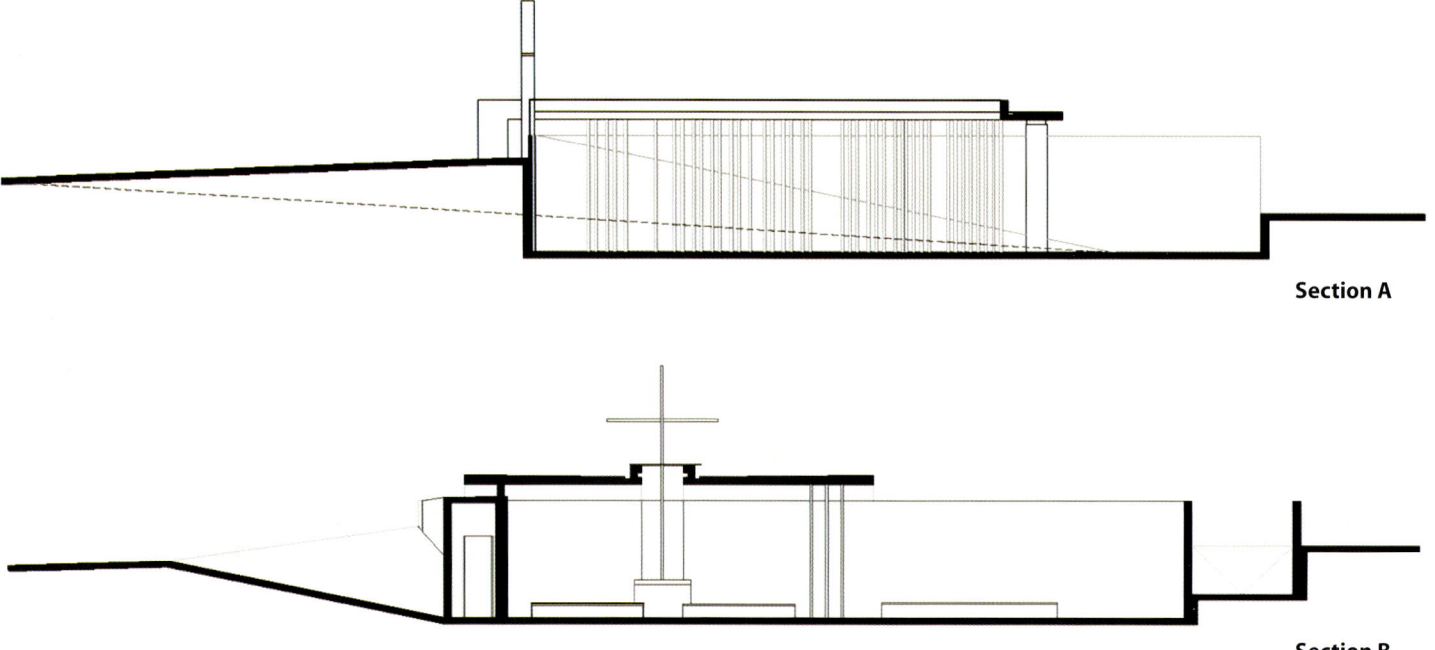

Section A

Section B

This is a place of faith where the light is embodied in forms that contain, transform, assimilate, and emanate a composition of dualities in search of the perfect balance; The corporeal and the ethereal accompany each other to achieve a speech where the material and the spiritual are harmonized to form a single space of art and seclusion.

Ecole Maternelle Antoine Beille

Architects: MDR Architectes
Location: Nissan-lez-Enserune, France
Area: 2631.0 sqm
Project Year: 2015
Photographs: Mathieu Ducros

Project Team: Sancie Matte, Frédéric Devaux, Arnaud Rousseau
Collaborator: Frédéric Ganichot

The project boasts a beautiful domineering location: it stretches to the South with an exceptional view on the entire Beziersr agglomeration below and its hilly green scenery and to the North on the Roman Oppidum, making Nissan one of the most renowned archeological sites in France.

Our priority was to enable children and school personnel to enjoy this beautiful view. In the meantime, the goal was to design buildings that could constitute a shield against the uncomfortable Northwest winds.

These data, added to the specificity of the program, the constraint of creating a passage between the elementary school and kindergarten while keeping some land on the western front for future extension, oriented this architectural design in long East-West layers: resulting in an optimal, even furtive implantation within the grounds. The landscaped public parking is situated at the forefront of the project on the Eastern side where the main entrance to the school and its atrium represent major public equipment for the commune, itself staged from the main entrance coming from the heart of the village.

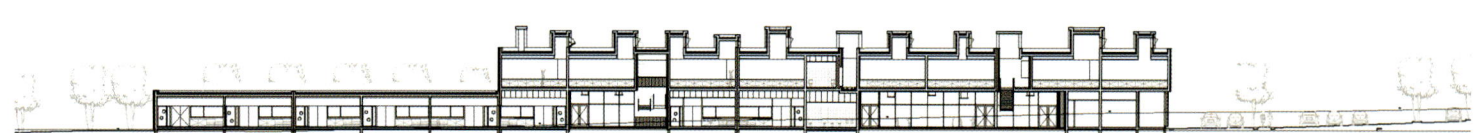

Longitudinal Section

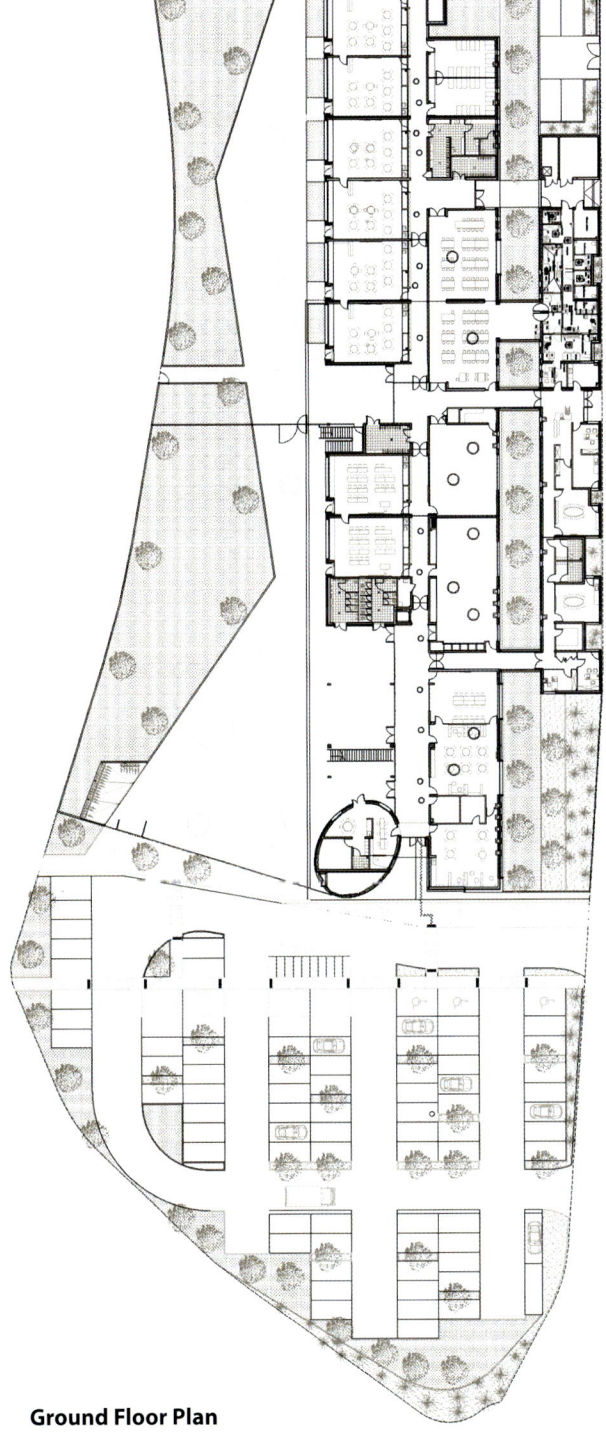

Ground Floor Plan

Common spaces consisting of the leisure center, activity rooms – motor development room, multifunctional room, library -, food halls and sleeping quarters for the kindergarten school are located in the central layer of the project, since these common areas constitute a real link between both school and are thus at the heart of the school complex. Symbolically, these areas are much protected as convivial and get-together spaces and they benefit from a specific attention: the roof is punctuated of round lighting fixtures, referencing directly to the "Dolias" characterizing the Oppidum of Enserune.

For the symbol, they represent the keeping of knowledge, for the aesthetic they bring a playful touch and colored lighting appropriated to these spaces; for the technical aspect they are a precious source of natural light completing the generous opening onto the patios, and those fitted in transparency as well as visual links with the playgrounds and traffic areas. They are located on the landscaped roof.

At ground level, the kindergarten classrooms open directly onto their playground, with for each one a dedicated educational garden. Inside, the access to the classrooms is extremely simple and visible via a vertebral circulation, in the continuity of the welcome hall: each class is color-coded immediately identifiable by smaller children.

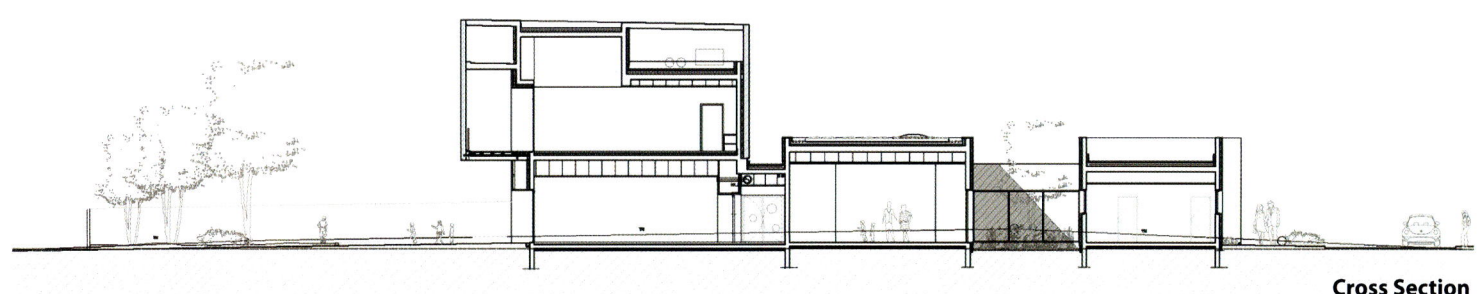
Cross Section

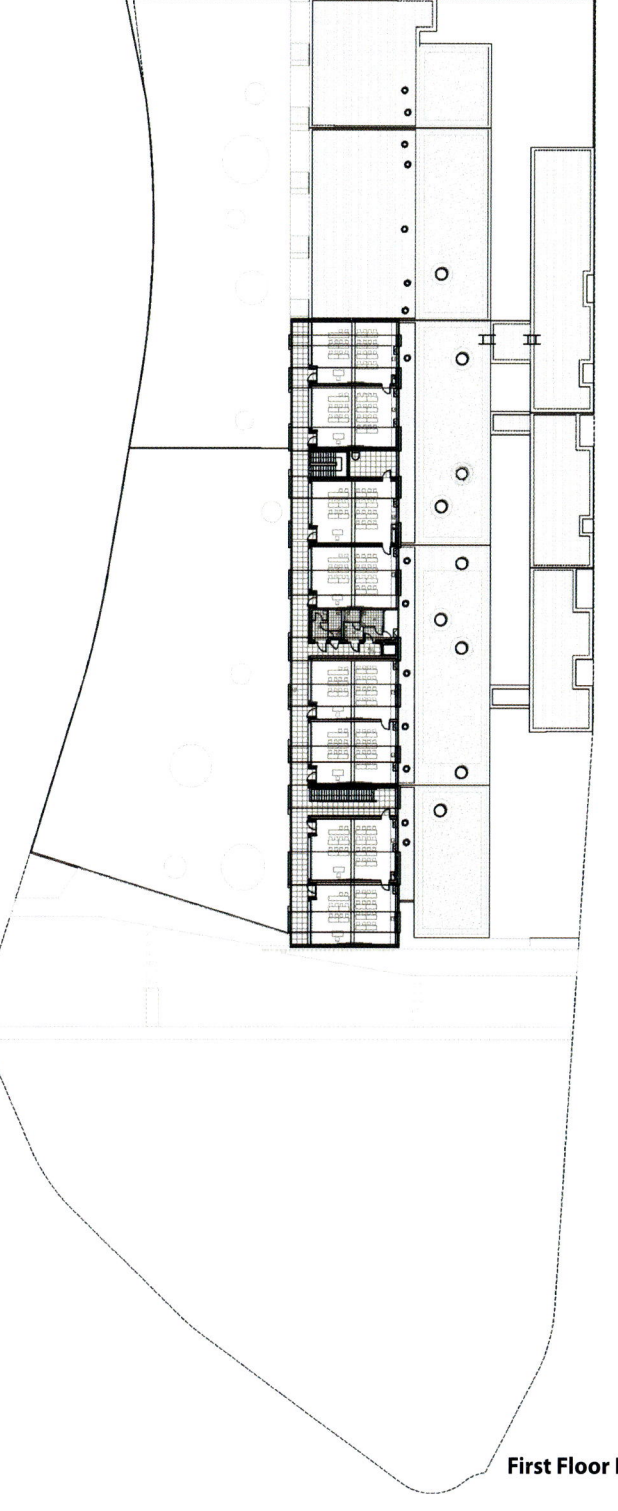
First Floor Plan

Box House II

Architects: Massive Order
Location: Kuwait
Architect in Charge: Muhannad Al-Baqshi
Design Team: Reem Aljalal
Area: 500.0 sqm

Project Year: 2015
Photographs: Nelson Garrido
Project Manager: Faisal Al-Hawaj
Construction Manager: Hamad Hussain
Construction: Massive Order

At its core, Box House II has a courtyard layout, with a living space on one side and sleeping quarter on the other side. This courtyard layout is placed in the first floor sandwiched between two other floors. The layout, then, disintegrates along a diagonal line dividing it into a large outdoor terrace and an indoor double-height common space with a prominent staircase.

The main facades of the house are composed as a single carved box. The two main facades conceal the internal logic of the plan behind large abstracted forms as to bring an element of surprise to the experience of the plan and section. This abstraction is especially legible at the middle section of the facades where the plan is introverted toward the courtyard leaving the façade clear of openings.

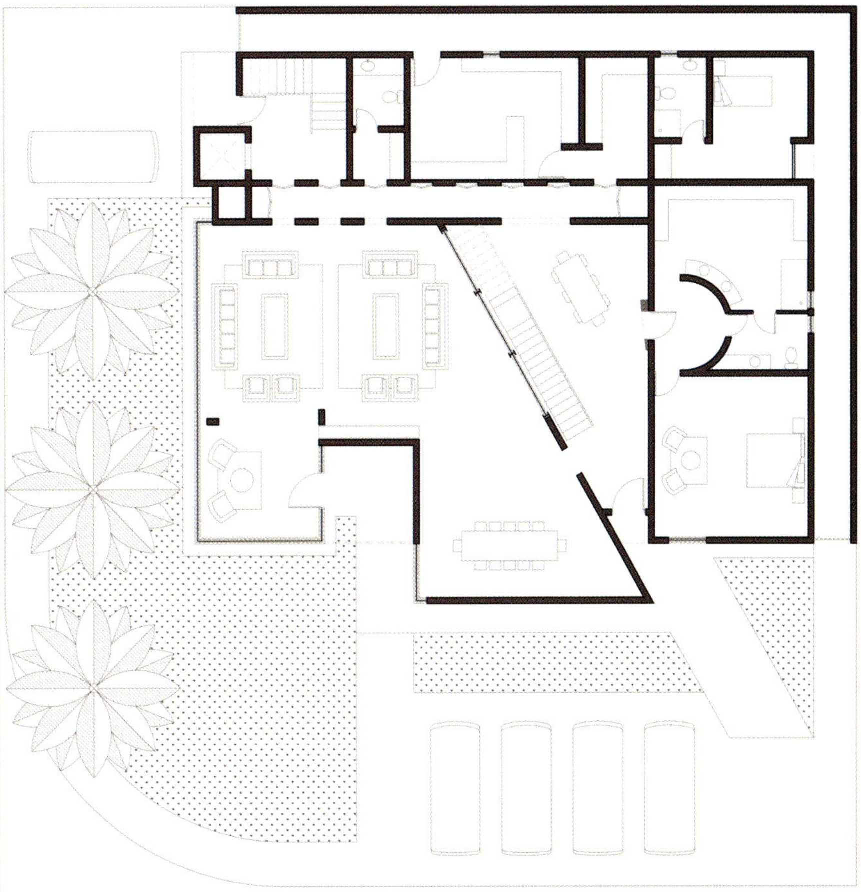

Ground Floor Plan

116

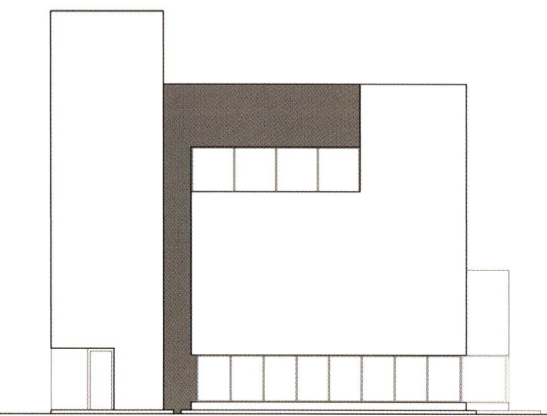
Elevation 1

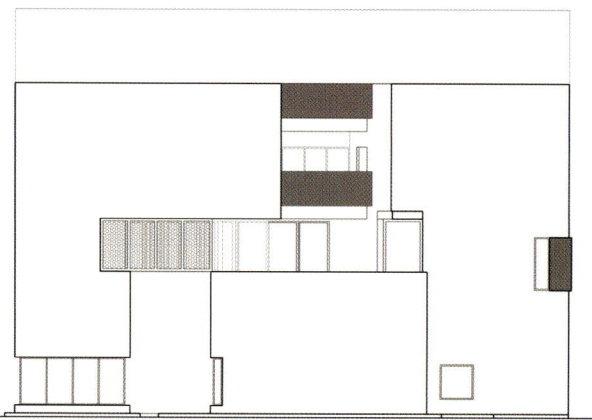
Elevation 2

Box House II modernizes the courtyard house order. It takes the archetype of rooms around an open space of the courtyard layout and reintroduces it in a sectional experience that cuts through all the houses' three floors. Then, it masks this experience with abstracted facades that leaves most of the delight to the users' daily lives.

A large cylindrical cut pierces through the second/third floor allowing the sunlight to penetrate into the terrace and the double-height space while delineating the functions in the upper floor. It, also, adds a dynamic sectional experience at the heart of the users' daily routine. The sleeping quarter, the living section, the main dining space, and the daily entrance have a direct relation to this sectional experience making it the heart of this design.

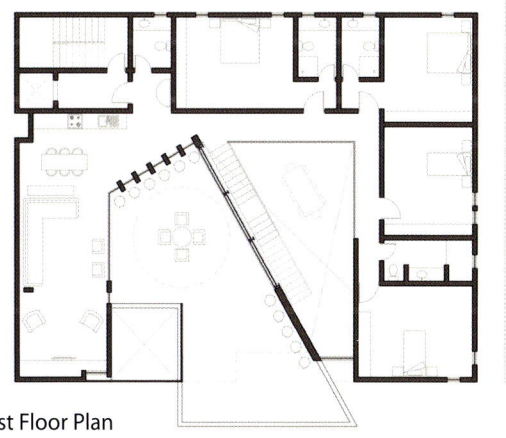

First Floor Plan

Building for 21 Houses

Architects: Roman y Canivell
Location: Seville, Sevilla, Spain
Project Year: 2015
Photographs: Miguel de Guzmán
Developer: Metrovacesa   **Contractor:** Vias y Construcciones, Grupo ACS

The plot is located in the city of Seville, in the scope of its historic centre and particularly in the sector of Murillo Gardens, in San Bernardo neighbourhood. The street facade has northeast orientation.

The urban environment in which we find our building is the typical historic house of Seville. This type has three floors often. In front of it is the City Hall of Provincial of Seville. We propose a project with a modern design but according to the characteristic buildings in the area. A premise is to "integrate" a contemporary architecture with a consolidated historic site. It is a multi-family building with garages and storerooms. It is consists of ground, first, second floor and penthouse. There are 21 apartments with 1, 2, 3 and 4 bedrooms, spread around a large central courtyard that is shaped as the star of the project.

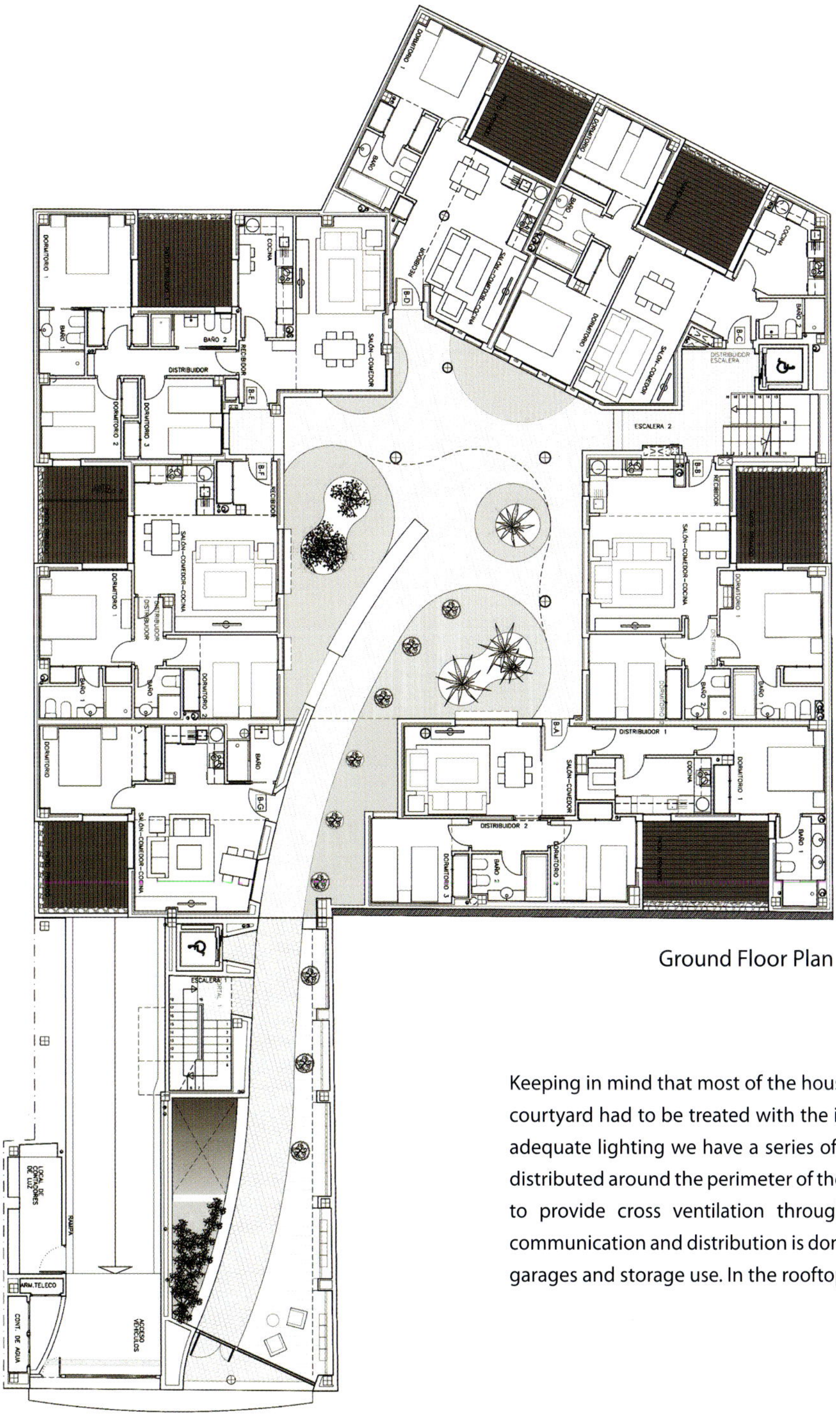

Ground Floor Plan

Croquis 1

Keeping in mind that most of the houses are inside, we understood that the design of the main courtyard had to be treated with the importance of exterior facade. For proper ventilation and adequate lighting we have a series of secondary courtyards that are attached to medians and distributed around the perimeter of the site. These courts contribute, together with the principal, to provide cross ventilation throughout the promotion. There are two cores of vertical communication and distribution is done through galleries. It also has two underground floors for garages and storage use. In the rooftop there is a small community pool.

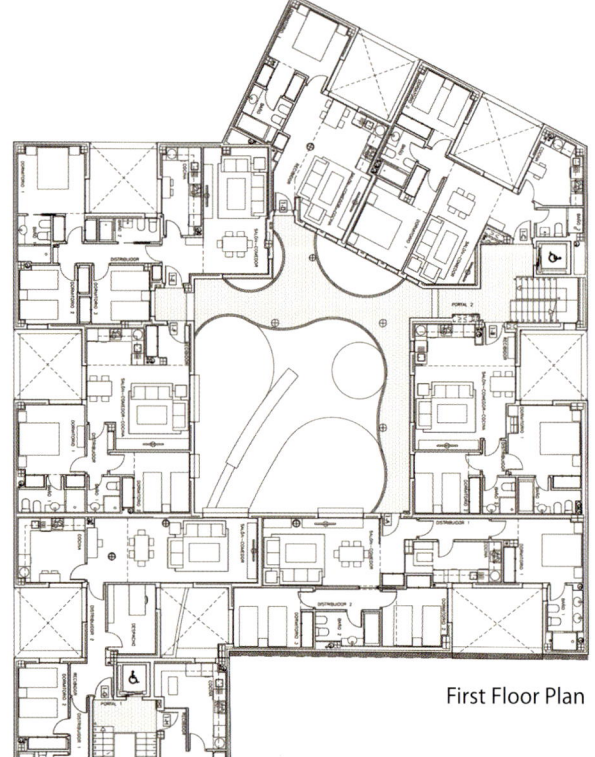

First Floor Plan

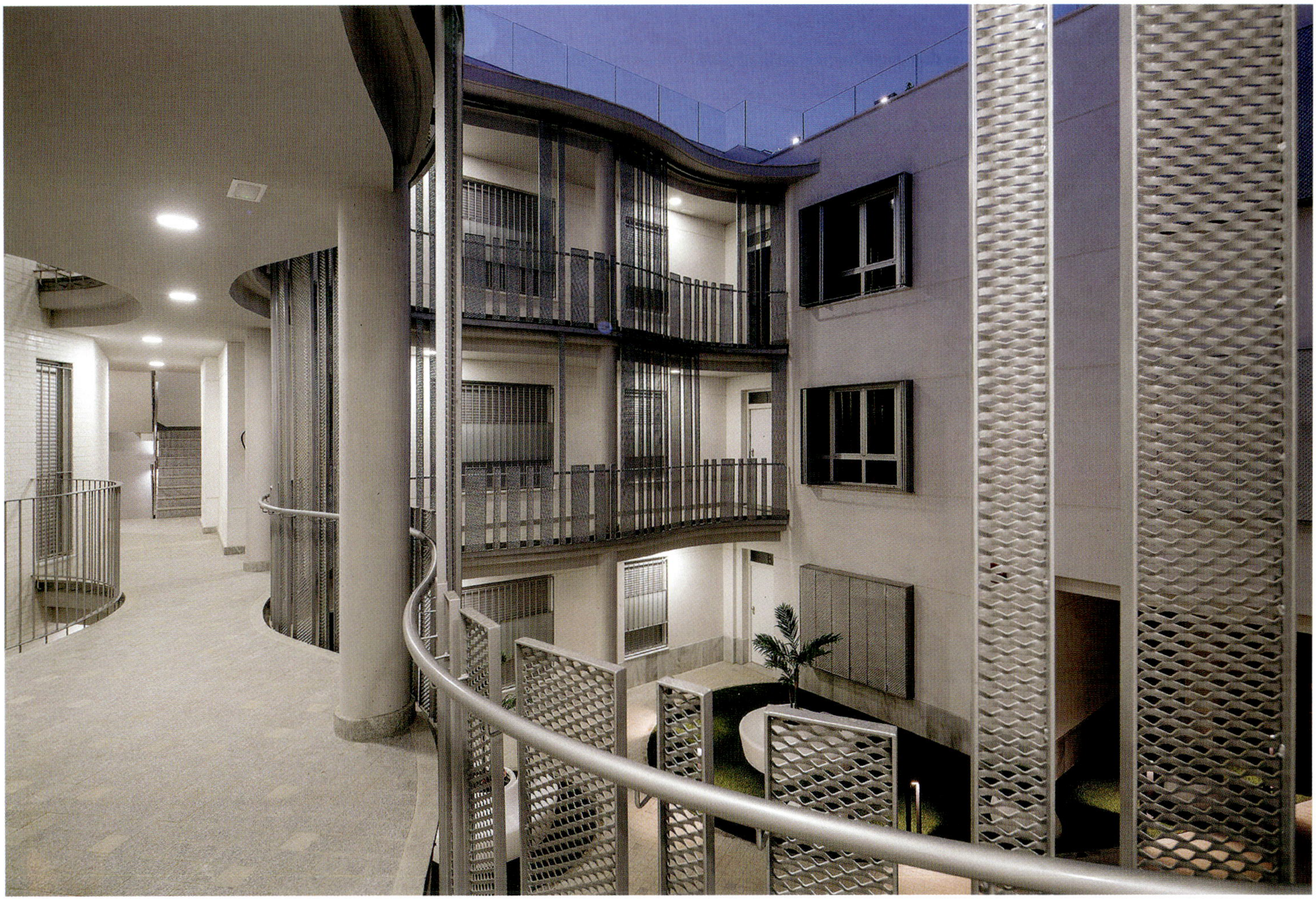

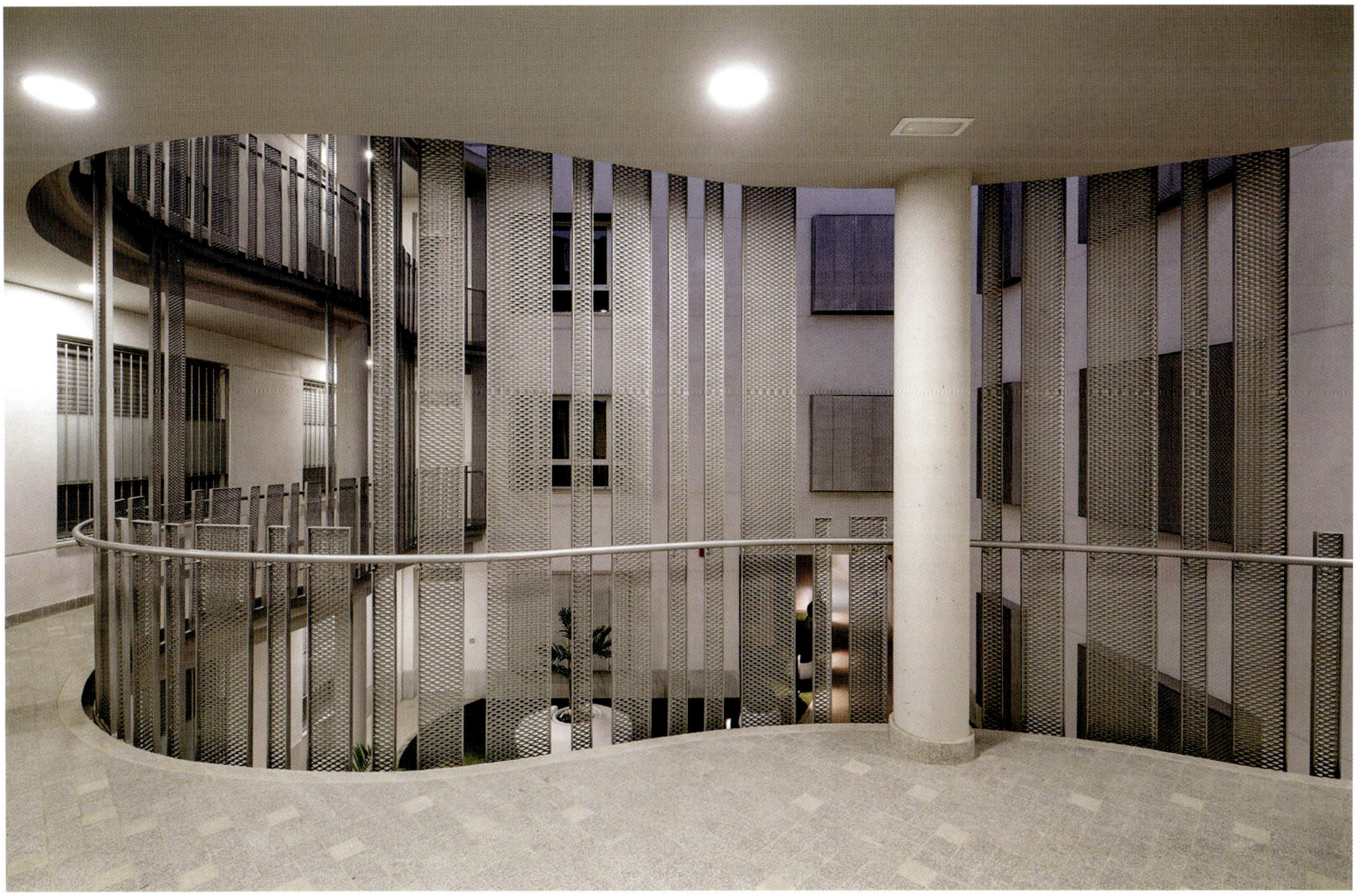

Algard Church

Architects: Link Arkitektur
Location: Rettedalen 7, 4330 Algard, Norway
Area: 1980.0 sqm
Project Year: 2015
Photographs: Hundven Clements Photography

Algard, just outside Stavanger, is a very vibrant congregation With many activities. One of the Aims of the new church was to bring together as many functions as possible under one roof. A challenge THEREFORE With the project was to create rooms suitable for a classroom, an office, a cafe, without Affecting the current Church space. By lowering the ground floor partially into the terrain, suitable space to Accommodate the range of functions room was generated, leaving undisturbed the sacral space on the upper floor. - A logical and functional solution That Makes the sanctuary natural and prominent center.

Another requirement from the congregation was the church That Should Have an iconic character and a clear Christian style. This request was solved by the buildings dynamic sculptural form, giving the appearance of single expression evolving out of the landscape, the sky Whilst Extending towards mirroring the shape and slope of the surrounding terrain.

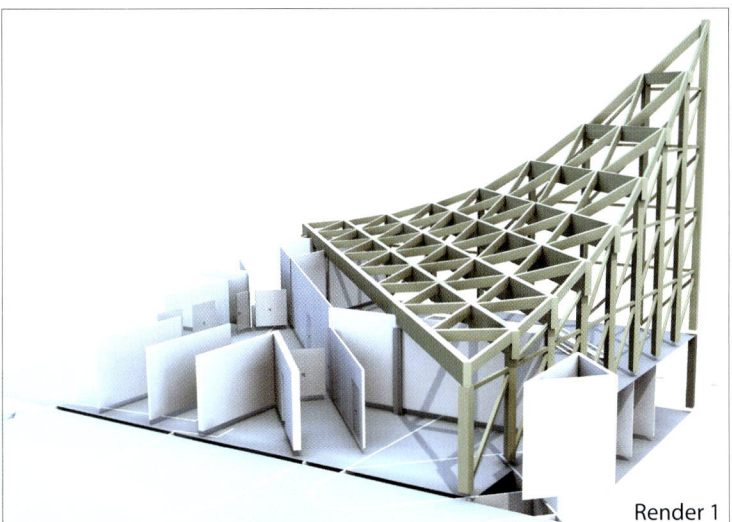

Render 1

Render 2

Render 3

Render 4

The building's shape defines the roof structure as a modern interpretation of a traditional church vault. The main structure Consists of Glulam beams in a network of triangles, as exposed in the Church room. Every other triangle is tilted upwards and Reflects the light into the church. The result is an optical refraction is seen in traditional dome vaults.

The closed triangles on the other hand, are designed with built-in LED lights. Additional daylight is in the church room is enhanced via the northern façade, Where the corner point is lifted slightly upwards in order to allow the light to fall along the floor in the back part of the room. Simple glass columns Provides additional daylight on Either Side of the building's high altarpiece in the glass. The landscape design merges the church into the surrounding countryside. A park Separates the form the church cemetery. The area can be an amfiteatreor Equipped with a suitable stage for concerts, sports and other activities.

The Church is designed in Accordance With The requirements for universal design.

Asahicho Clinic

Architects: hkl studio
Location: Chiba, Chiba Prefecture, Japan
Area: 310.0 sqm   **Project Year:** 2015
Photographs: Shinkenchiku-sha, Tetsu Hiraga
Site Area: 333.48 sqm

Signage Designer: TOKOLO.COM / Asao Tokolo, BANANA DESIGN
Artwork Artist: TOKOLO.COM / Asao Tokolo
Main Structure System: RC
Collaborating Architect: Michio Kinoshita / Workshop

The clinic is located in a residential area near Tokyo, where currently there is an increased elderly population. This project is intended to establish an "open" clinic for the local area. Which is different from the conventional hospital. The Idea is, that in the future the residents can visit their family doctor in this building, as opposed to the specialized medical service in the university hospital.

To have the benefit of the maximum area on the following site boundary, the floor plan results as L-shaped. The building is northeast oriented, opening through a dynamic facade with several recesses that allow to let the natural light come inside, while at the same time keeping the privacy of visitors. This concept of recesses creates a strong relation between the external natural environment and the interior space. That, at the same time, creates a variety of niches that give the patient the possibility to choose their favourite one.

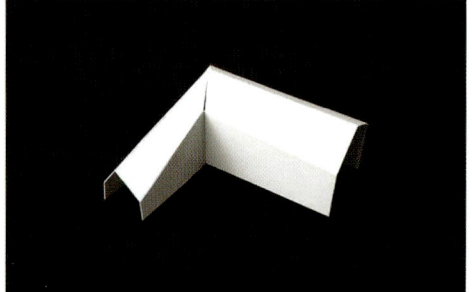

The roof gets gradually higher until the center of the building.

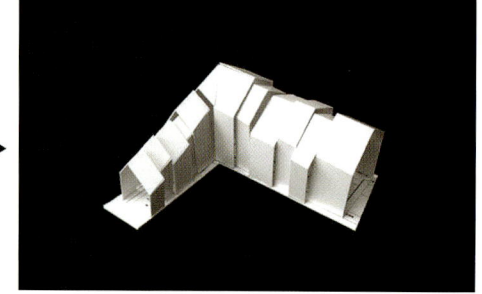

Sliding several parts of building volume makes the opening.

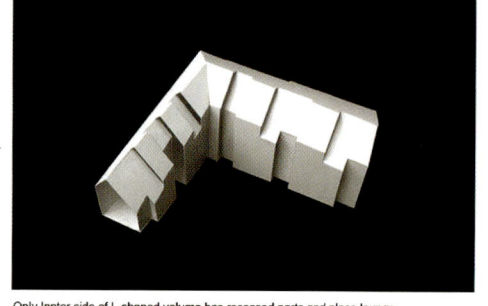

Only Innter side of L-shaped volume has recessed parts and place lounge. Keeping outer side of L-shaped volume flat makes a stable structure.

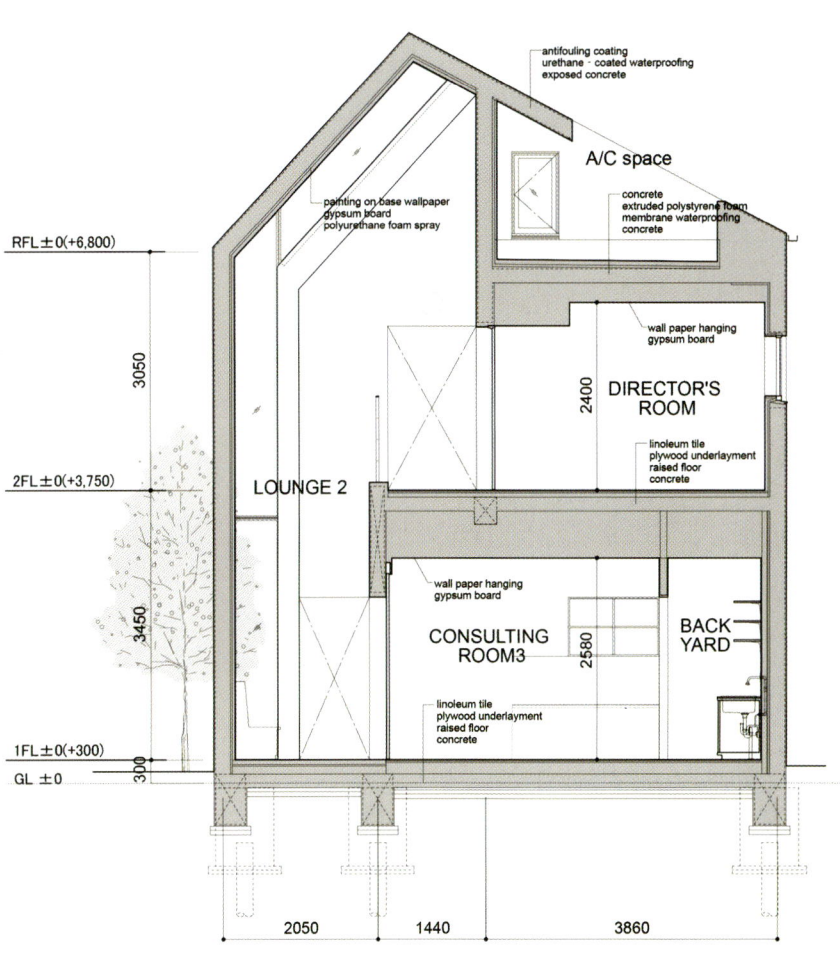

C-C' SECTION S=1:100

The building is therefore composed of two floors. On the 1st floor we placed all clinical examination rooms to facilitate the elderly and disabled people, and the service spaces for the staff members are situated at the 2nd floor.

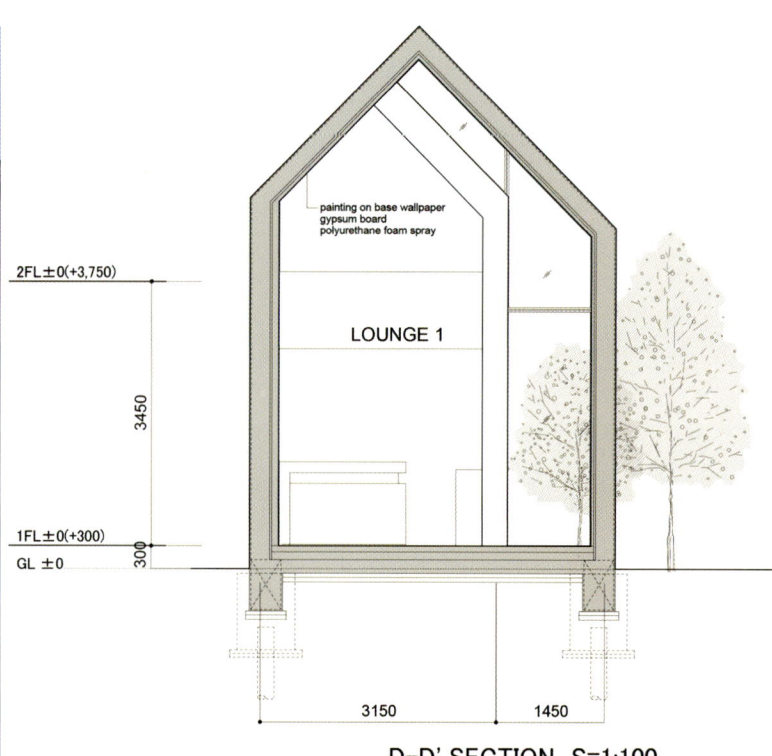

D-D' SECTION S=1:100

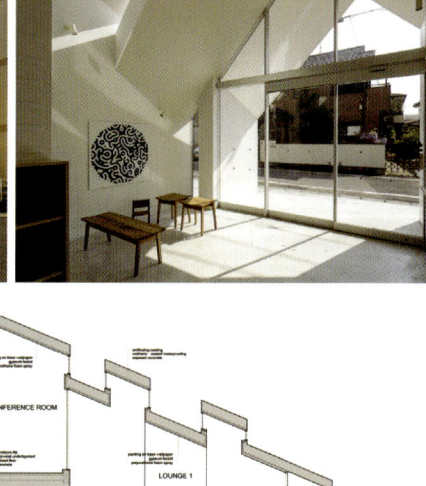

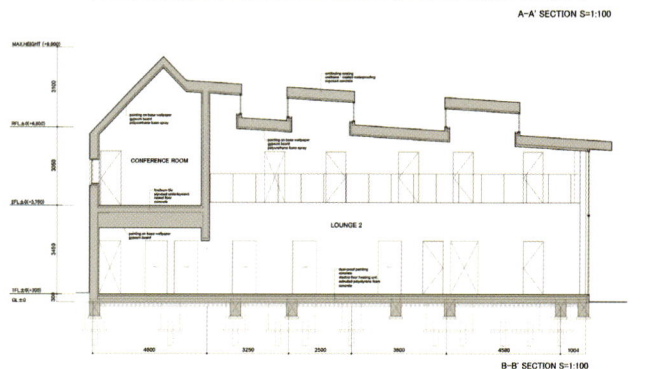

A-A' SECTION S=1:100

B-B' SECTION S=1:100

The roof gets gradually higher until the center of the building. This transition produces spaces with right proportions, from a cozy entrance to a double height space for waiting, which reduce the feeling of pressure for the surrounding environment. For keeping the space more open at Lounge2, we decided to set back the wall at the 2nd floor. To realize this geometric shape we decide to use a RC structure. The West part is made of a composition of many wall-type RC structures, instead the East part is composed of series of single complex elements (walls and roof are one element) with a wall-type frame RC structure.

At the construction process, we decided to use "cut" plywood framework as the shape of the "unrolled" 3D model, such as paper model. This precise work during the design phase allowed us to achieve an exact geometric shape with high-accuracy technic. We hope this clinic will surpass the general function of the normal clinic and develop a close relationship with the local residents, as the "house" of the quarter.

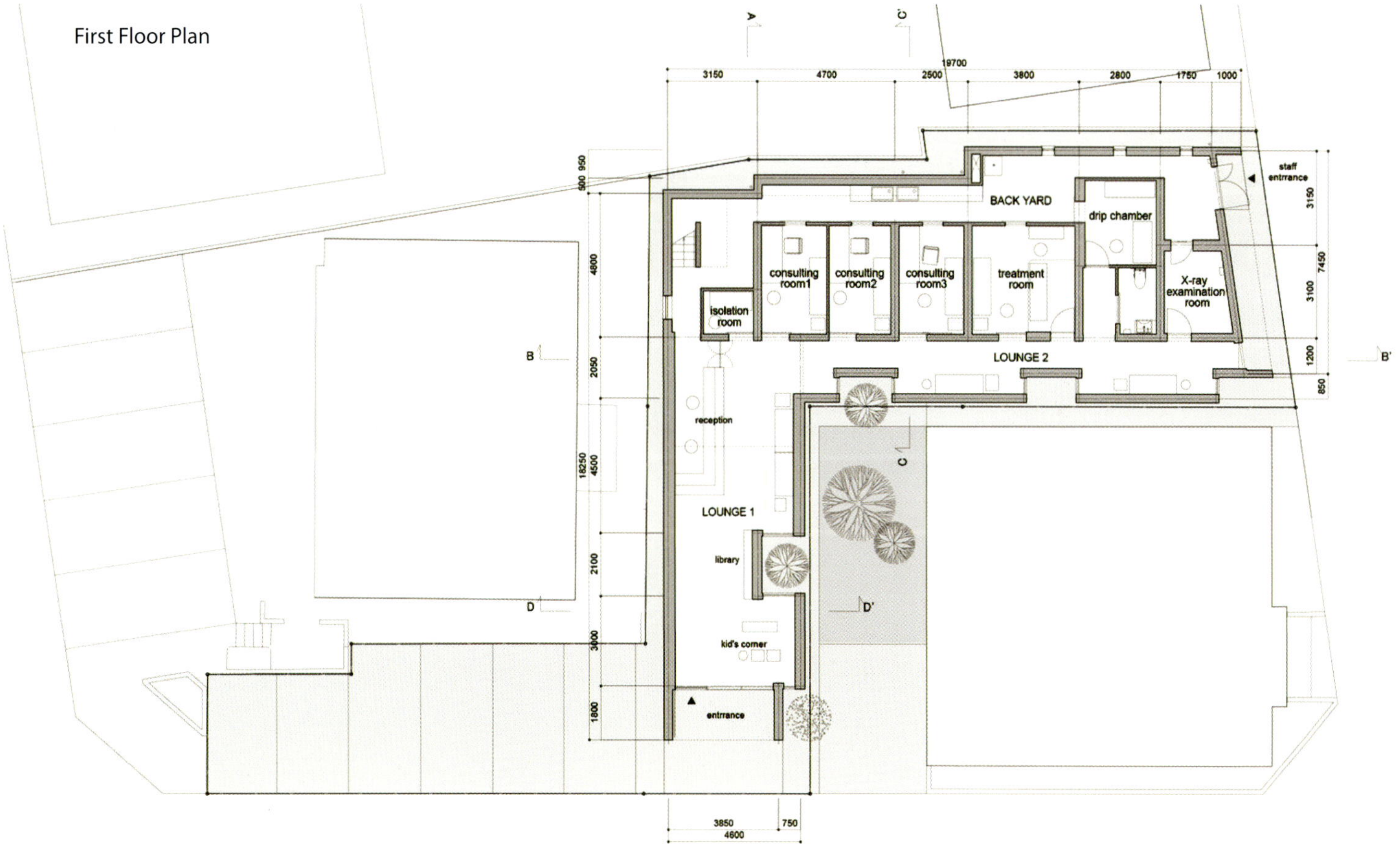

First Floor Plan

22 Greenleaf Place

Architects: IX Architects
Location: Greenleaf Pl, Singapore
Design Team: Chu Yang Keng & Chin Yew Chong
Area: 323.0 sqm   **Project Year:** 2015
Photographs: Lim Jing Qian

Principle Architect: Chu Yang Keng
Builder: EKHL Pte Ltd
Structural Engineer: TnJ Consultants LLP
M&E Engineer: BS Ong Engineering Services
Quantity Surveyor: QS Consultants Pte Ltd

Located at Bukit Timah within a quiet neighborhood, 22 Greenleaf Place presented IX Architects with the perfect opportunity to execute & realize the firm's design philosophy and beliefs. Approached by the client to design and reconstruct their existing semi-detached house, IX Architects began the design process by keeping in mind the client's brief to achieve a modern yet timeless building form.

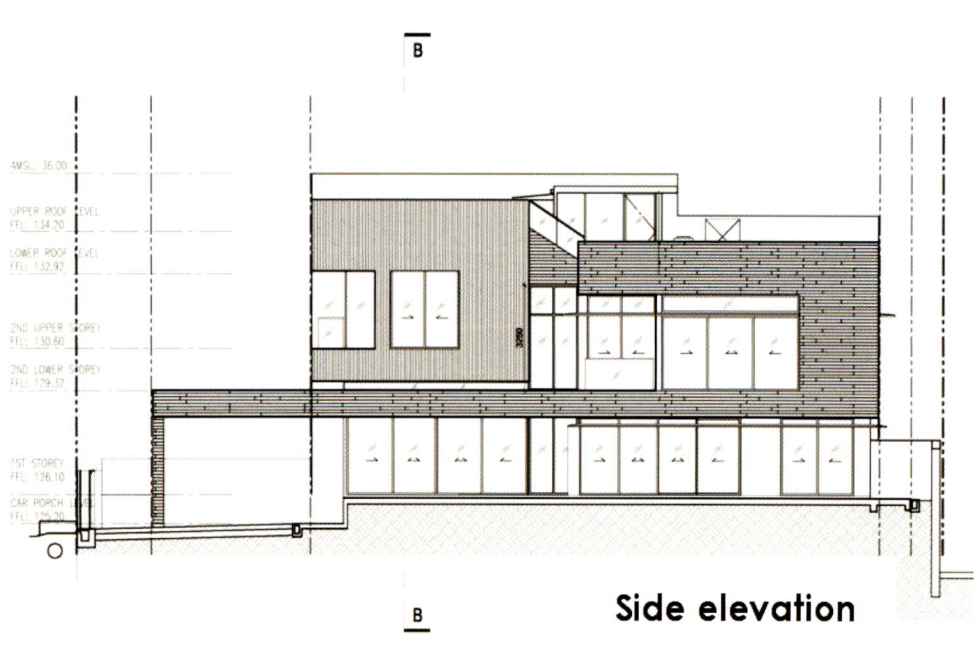

Side elevation

Functionally, the spaces are classified into 3 main zones; living & entertainment, service areas, and the bedrooms. These zones are also reflected in the building form – a "transparent box" on the 1st storey suggesting common spaces, more enclosed spaces at the rear corner of the 1st storey indicating service areas and the celebrated "solid boxes" on the 2nd storey implying more private spaces.

Section A-A

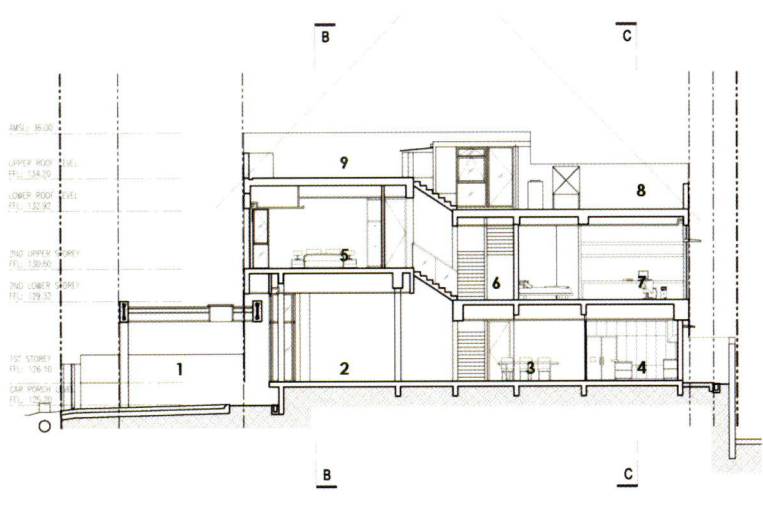

1. Car porch
2. Living
3. Dining
4. Kitchen
5. Bedroom
6. Lounge
7. Master bedroom
8. Roof deck (upper)
9. Roof deck (lower)

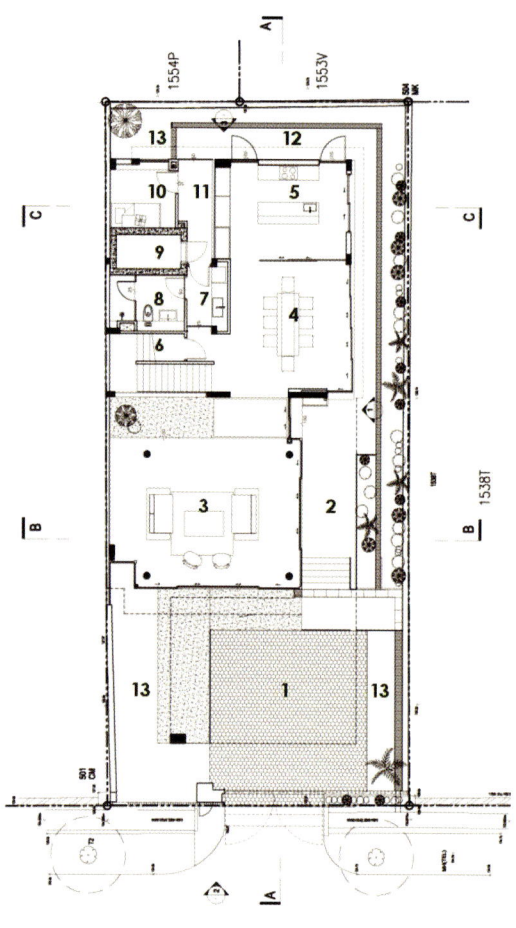

1st storey plan

1. Car porch
2. Outdoor foyer
3. Living
4. Dining
5. Kitchen
6. Store
7. Laundry
8. Powder room
9. Household shelter
10. Maids room
11. Service yard
12. Back yard
13. Landscaping

Living spaces on the ground level are visually & physically connected to the external landscaping. Full height glass doors can be slide away to allow effective cross ventilation and for activities to spill out to the external foyer and landscaped spaces. The 1st storey is kept at the existing elevated level which is higher than the surrounding roads. This has allowed the living & dining rooms at 1st storey to better capture the passing breezes providing natural cooling to the internal spaces.

The front and rear wings housing the 2nd storey bedrooms are distinctly identified as 2 separate boxes. The white (front) box in textured grooved finish and the grey (rear) box express the concrete walls in their raw form. A central staircase links all the 3 main zones and allows visitors direct access from the ground floor to the rooftop terrace. The rooftop terrace, where family gatherings & barbeques can take place provides unblocked scenic views towards the city in which the owners were deprived of in the original house.

The existing staircase structure from 1st to 2nd storey was retained and given new life with the introduction of new finishes to the floor and wall. The staircase walls and handrails were detailed to incorporate back lighting and light fixtures illuminating the vertical circulation area at night. Naturally lighting filters in through the central skylight and air well such that practically all the common areas are naturally lit ted during the day..

The master bedroom suite on the 2nd level is designed with a walk-in wardrobe and a spacious master bathroom. Natural lighting floods the bathroom space with the use of full height glass sliding windows with adjustable louvers screens for privacy.

170 Amsterdam

Architects: Handel Architects
Location: 170 Amsterdam Avenue, New York, NY 10023, USA
Area: 21235.0 sqm   **Project Year:** 2015
Photographs: Bruce Damonte
Client: Equity Residential

Structure: DeSimone Consulting Engineers
Lighting: Clinard Design Studio
Exterior Wall: IBA Consulting & Engineering
Landscape: Blondie's Treehouse

Our client wanted a solution that maximized floor area on a long narrow site on the Upper West Side of Manhattan. They did not want an all glass building, but wanted the benefits of large windows without compromising energy efficiency. Creating a new terminus for W68th Street, 170 Amsterdam sits between Central Park to the east and the landscaped open space of the Lincoln Towers superblock to the west. The building's architecture and the exoskeleton that defines the exterior is derived from its location between these large green spaces and its immediate context. The concrete used to create the exoskeleton is the result of a specialized mix that gives the material the appearance of limestone, a nod to the buildings in the Lincoln Square neighborhood. The formwork for the round, crossed column shapes involved an intricate fiberglass system with multiple units that were tightly connected together to achieve maximum reuse and economy.

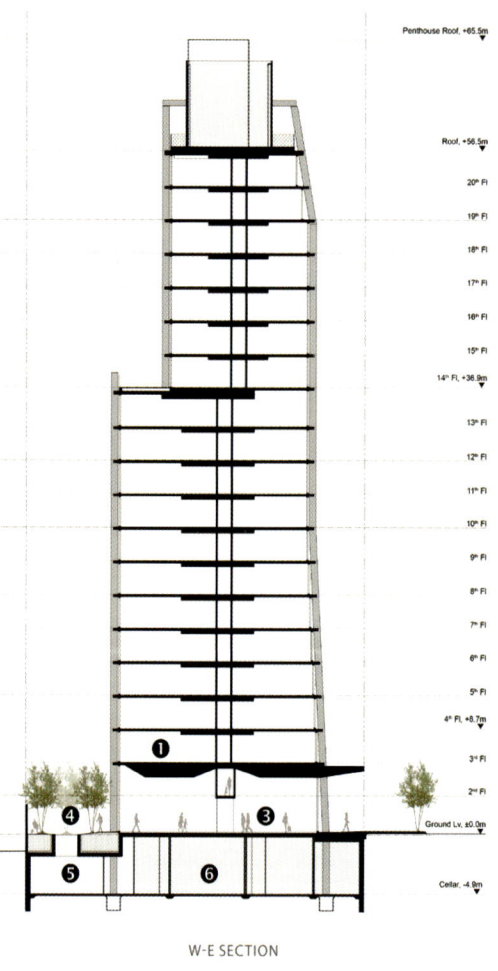

W-E SECTION

170 AMSTERDAM

STRUCTURAL REINFORCING
Vertical reinforcement bars limited to allow for intercrossing without congestion

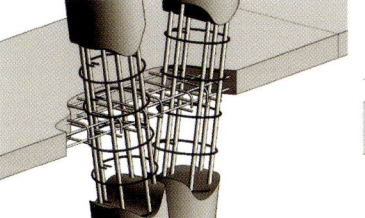

FORMWORK
Custom fiberglass forms reinforced with plywood ribs and connected together

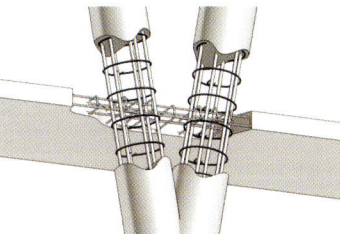
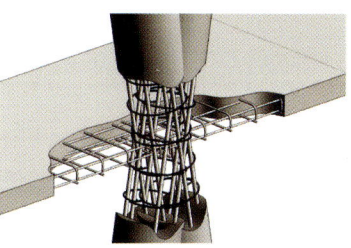

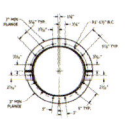
BOTTOM FLANGE

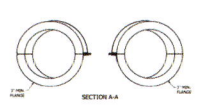
TOP FLANGE

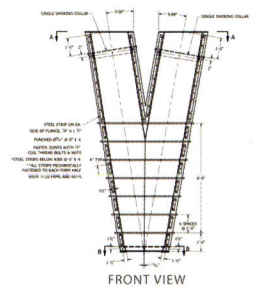
SIDE VIEW · FRONT VIEW

CONCRETE
The concrete mix was a dense, high fluid self-consolidating mix, with small evenly graded aggregate, using gray cement and slag to achieve a light gray finish and meet LEED standards. Using the small aggregate mix enabled the concrete to be placed into the blind areas of the formwork with no vibration. The concrete was pumped into the formwork with strict procedures for making sure there was no entrapped air to cause honeycombs or bug holes on the column surfaces.

The intersections of the structure rise to the top of the building at different heights, giving the appearance of a façade in motion while also allowing for the prefabricated fiberglass formwork to be reused with the concrete cycle. Moving the structure to the outside of the enclosure freed up valuable interior space that would have been occupied by columns and the projecting slabs and columns provide a veil over the all glass façade and act as a shading device. The deep façade of the diagrid connects with the muscular buildings of lower Amsterdam Avenue, Lincoln Center, the Upper West Side and iconic Chicago architecture. At ground level, the columns create a dynamic street wall, with the exposed structure angling into the sidewalk and piercing the solid form of the building canopy. At the top, the building's volume ends while the skeleton continues, creating a structural canopy for the rooftop spaces. Inside, the exposed concrete columns angle through the public spaces of the building piercing the floors and walls of the lobby, common rooms and corridors, and disappearing into the ceiling above. In the apartments, seeing the structure through the floor-to-ceiling glass has the effect of being suspended in a treehouse, held up by the branches of the building's exoskeleton.

170 AMSTERDAM

CORE & SLAB STRUCTURE

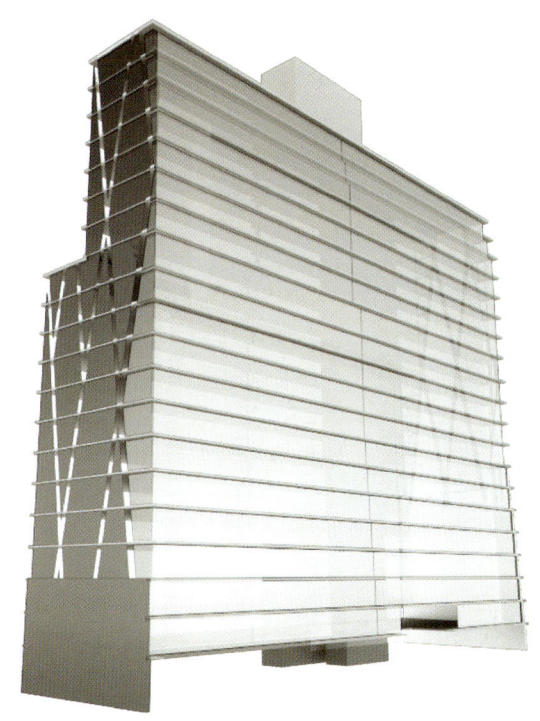

HIGH PERFORMANCE GLAZING

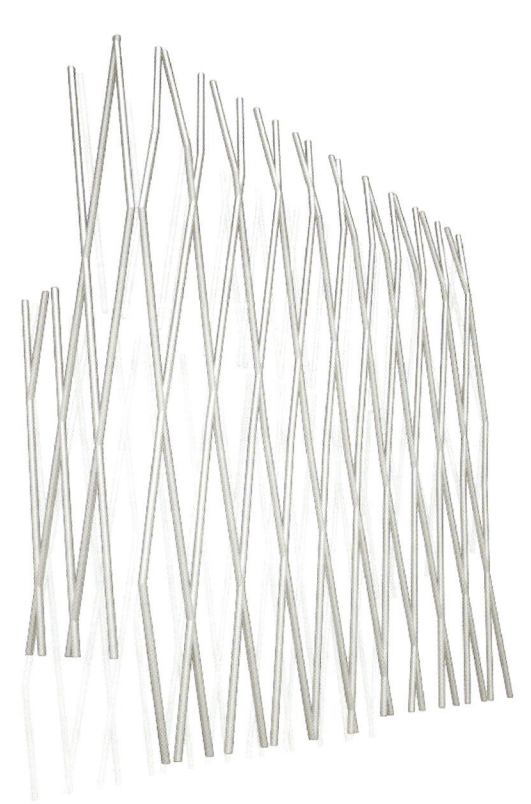

COLUMN STRUCTURE

EXOSKELETON

Abrantes Municipal Market

Architects: ARX Portugal
Location: Rua da Esplanada 1º Maio, 2200 Abrantes, Portugal
Area: 1280.0 sqm
Project Year: 2015
Photographs: Fernando Guerra | FG+SG

Design Team: José Mateus e Nuno Mateus c/ Ricardo Guerreiro, Fábio Cortês, Ana Fontes, Bruno Martins, Filipe Cardoso, João Dantas, Marc Anguill, Sofia Raposo, Miguel Torres

The Abrantes Municipal Market is located in the transition to the historical center, where the Rodoviária do Tejo's workshops once were, so badly in ruins that demolition was recommended. It is an urban lot between two streets, each at a different height: at the bottom (to the west) we have the Largo do Tribunal (Court Square); at the top to the east, the N. Sra da Conceição street. It is an extraordinarily narrow lot for the program at hand, and this feature has ultimately determined the design project.

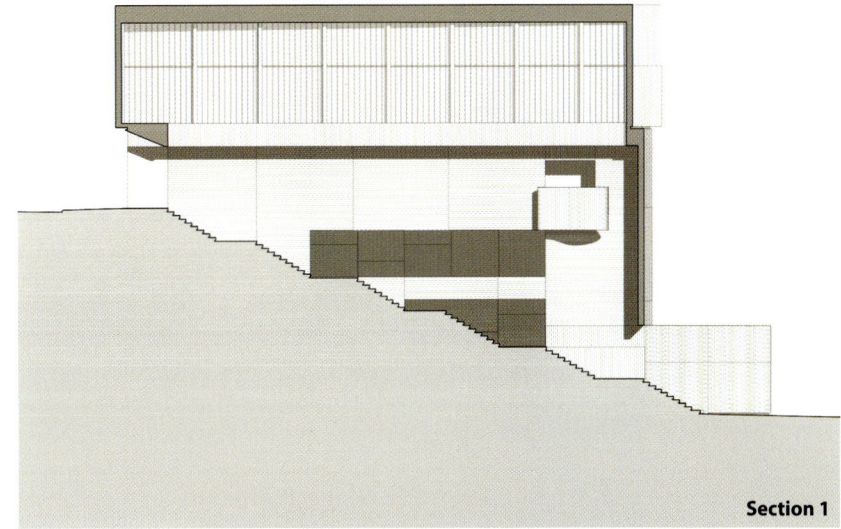

Section 1

The program stated the importance of connecting these two streets, thus creating an upwards route to the Museu Ibérico de Arqueologia e Arte (Iberian Museum of Archeology and Art), intended to be built in the S. Domingos Convent at the highest city level. Moreover, a closer analysis of the place revealed the need to consider the impact of this new building among the surrounding dwellings, diverse in architecture quality and construction dates, but also considerably smaller when seen from the west.

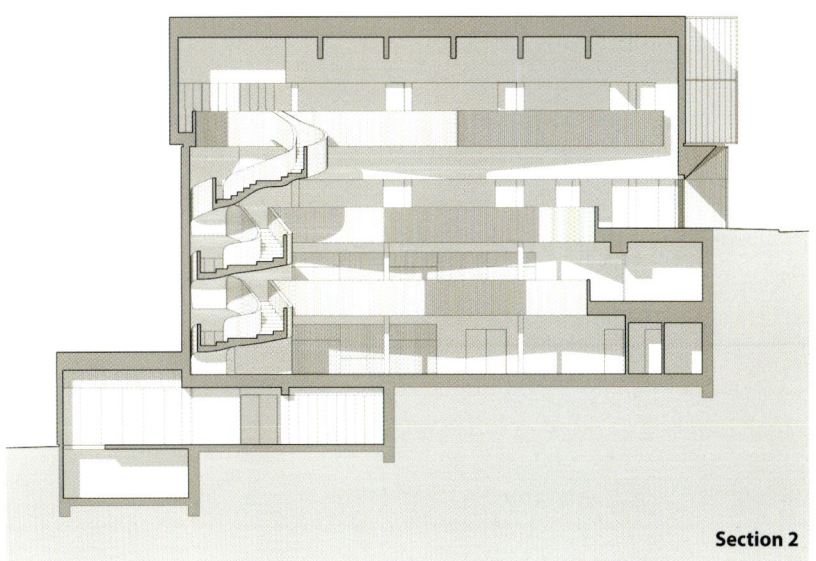

Section 2

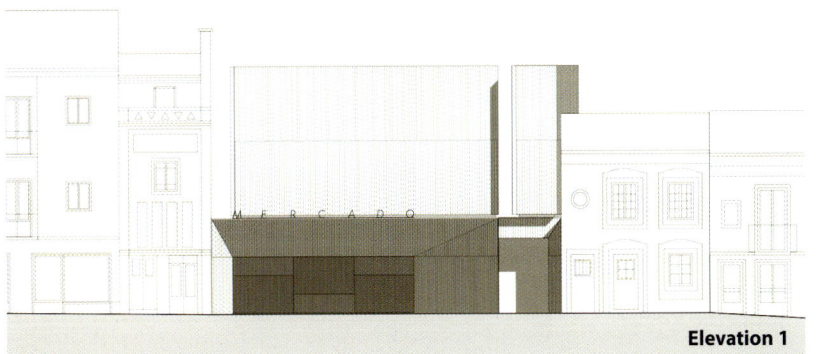

Elevation 1

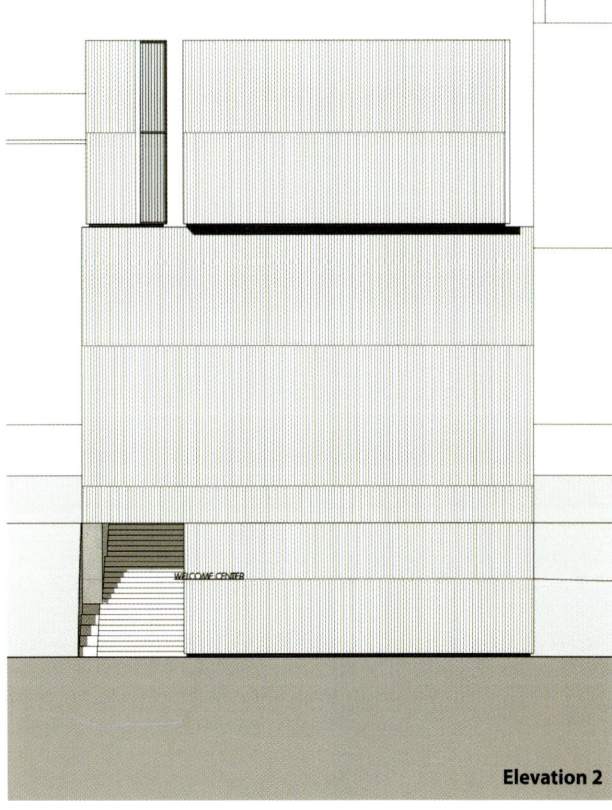

Elevation 2

Typology-wise, a market is a building where the concept of public space is taken to the limit. Still present in so many cultures, the market takes place on the streets, and sellers do their businesses in carts, makeshift stands and tents, making the market place and the city space merge or coincide.

The new Abrantes Market is at once building and street. One can cut across, going from one street to the other, either through the stairway opened at the northern edge, or wandering between the stands and the spiral staircase located at the southern edge. Deep down, it is basically a street, shaped with walls and covered by a white-washed shell of exposed concrete. On the top of the building two volumes capture the sunlight that flows to the lower floors through openings in the slabs at each level, softly lighting all spaces, highlighting the concrete's texture and exposing the passing of the hours.

Townhouses with Private Courtyards

Architects: baan puripuri
Location: Bangkok, Thailand
Project Team: Pajitpong Pongsivapai, Khajorn Jaroonwanit
Area: 233.0 sqm   **Project Year:** 2015
Photographs: Beer Singnoi

After decades of development, many townhouses, especially in Bangkok, have turned, unintentionally, to dark and air-con ventilated box-type buildings. To bring back improve living quality, we create an air well and a pocket garden in the middle of each unit. This internal courtyard provides natural light & ventilation, green and relaxing space while maintaining privacy for inhabitants.

Internal balconies and key functions are placed and stacked next to the courtyard with a semi outdoor bridge linked these functions together. Landscape at the center is also the main focus point of the house with random brick walls behind as a backdrop.

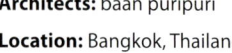

The front facade is series of random brick walls which designed as vertical fins shading double height living area from late morning glare and heat. Cantilevered planter boxes protruding from the facade add a touch of nature for both neighbourhood and residences.

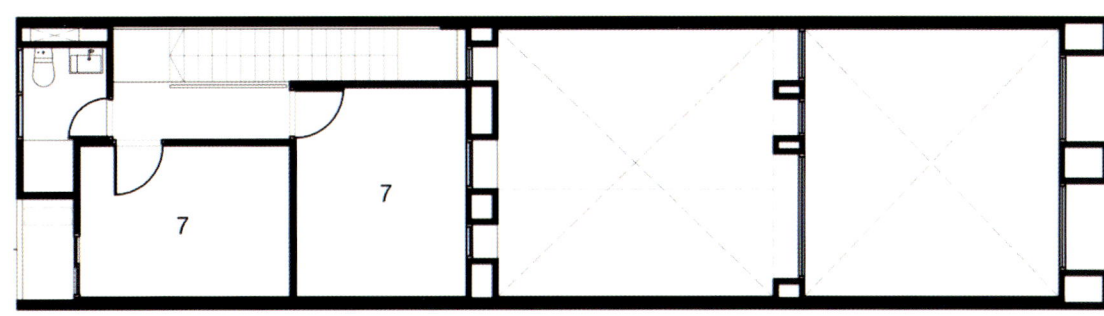

3rd FLOOR PLAN

1. Parking
2. Landscape
3. Terrace
4. Dining room
5. Kitchen
6. Living room
7. Bedroom

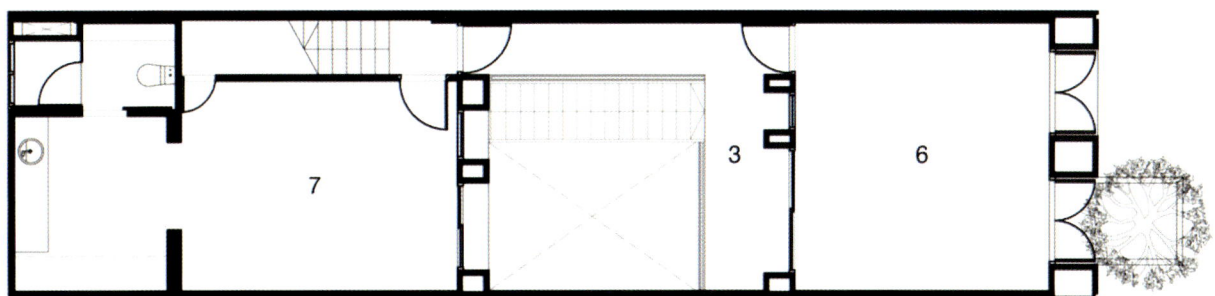

2nd FLOOR PLAN

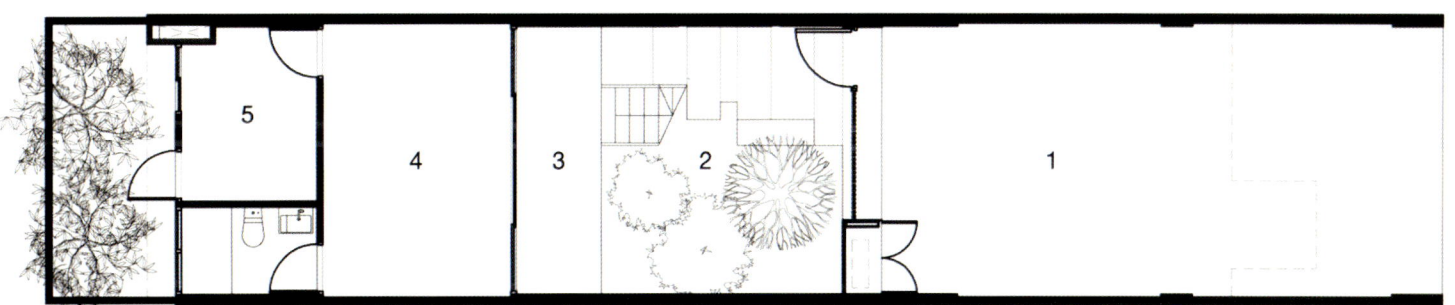

GROUND FLOOR PLAN

The Europe - Far East Gallery

Architects: Ingarden & Ewý Architekci
Location: Kraków, Poland
Area: 2675.0 sqm
Project Year: 2015
Photographs: Krzysztof Ingarden

Interior Design: Ingarden & Ewý Architekci
Landscaping Design: Land-Arch / Karolina Bober, Małgorzata Tujko
Road Design: Zdzisław Krzysztoforski
Architectural Design: Ingarden & Ewý Architekci
Co-designer: Jacek Ewý

The idea for the new building came up in 2004 in connection with the Manggha Museum's acquisition of the small plot of land adjacent to its site, along ul. Konopnickiej. The project initiators' main objective was to broaden the scope of its intercultural artistic and exhibition activities – from Polish-Japanese to European-Far Eastern – which coincided with Poland's accession to the European Union. The new exhibition space was intended for presentations of European art and the cultures of Southeast Asia, including the Indian Peninsula. The programme of the Gallery provides for shows of old and recent art. The new building has two exhibition rooms on two storeys. As intended by the management of the Manggha Museum, both are classic white cube spaces, as neutral in expression as possible, to allow for the display of various forms of mostly modern art.

Sketch

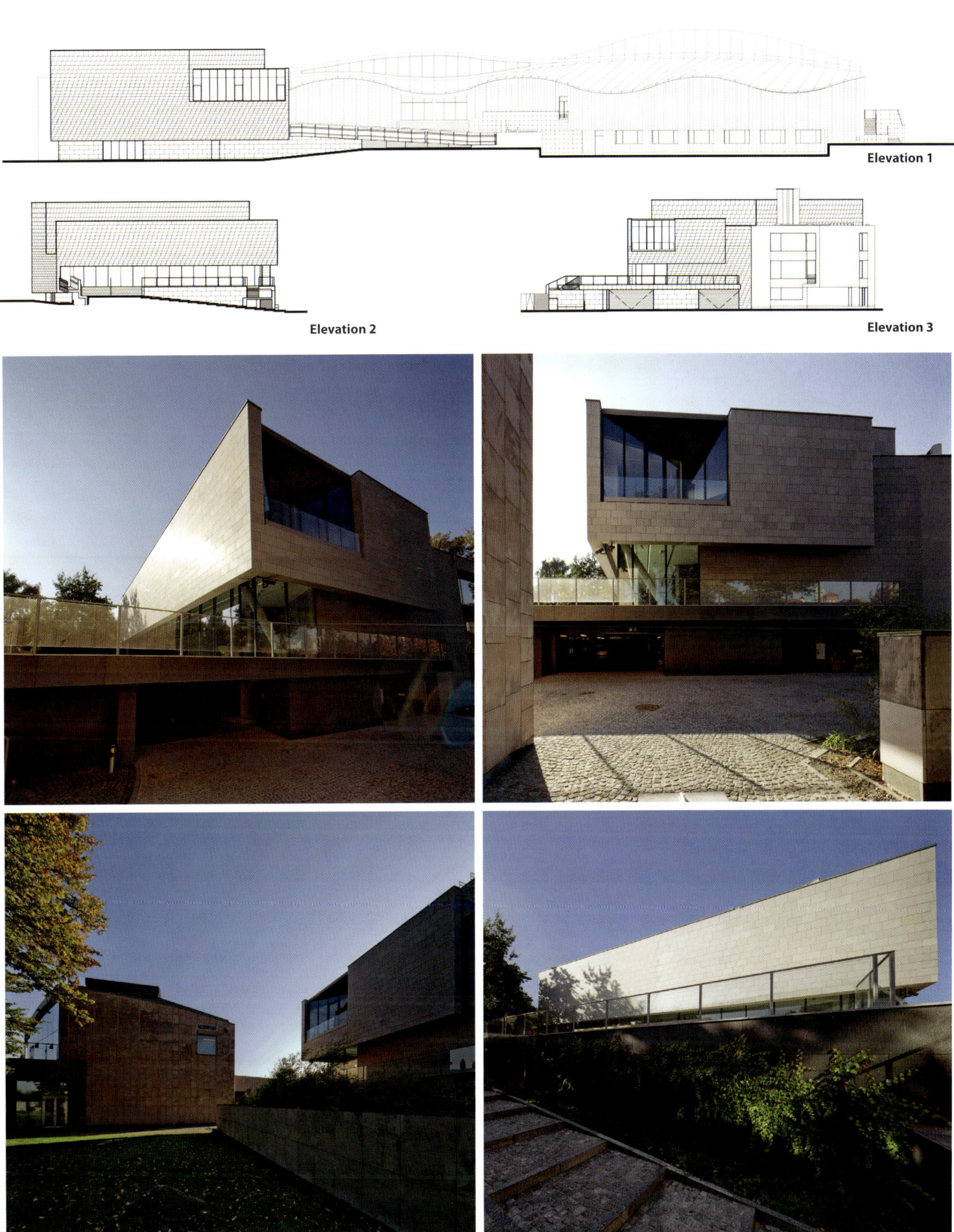

Elevation 1

Elevation 2

Elevation 3

152

Floor Plan 1

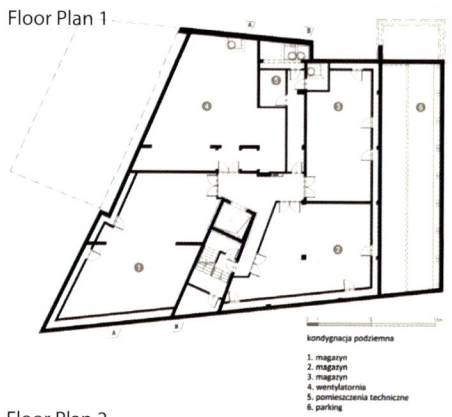

Floor Plan 2

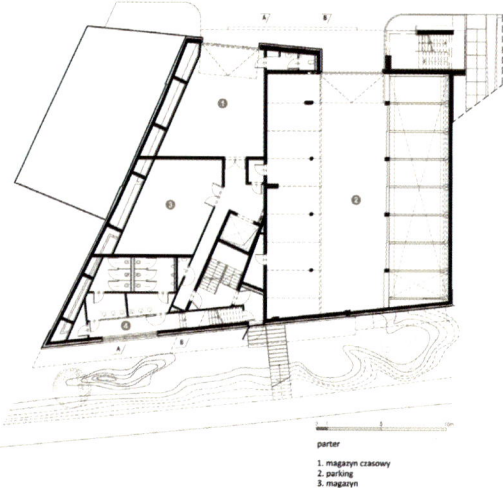

Floor Plan 3

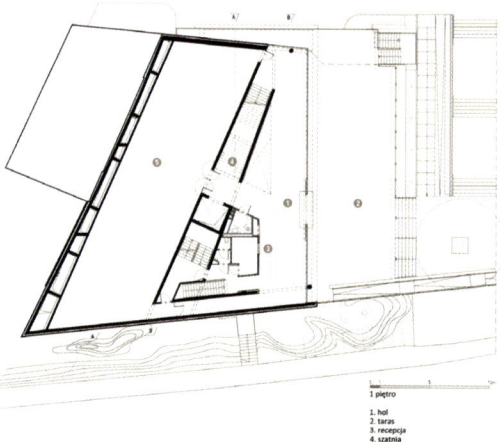

The Manggha building has keenly felt the lack of this kind of room because the characteristics of its main exhibition space were tailored to the requirements specific to the display of historic art, particularly paper, which called for the use of special showcases, dedicated lighting, and appropriately controlled humidity levels and temperatures. The narrow specification of those conditions limited the possibilities for exhibiting contemporary art. The new building is boosting the scope of potential displays considerably, allowing the curators much greater freedom in terms of use of space and types of exhibition projects. Furthermore, the Gallery is fitted with infrastructure required for the preparation of exhibitions, as well as storage space, offices for curators, and dedicated parking spaces inside the building.

The main building of the Manggha Museum is a complete, finished composition, with a sophisticated and unique roof form, characterized by a juxtaposition of several planes based geometrically on the hyperbolic paraboloid. Any action seeking to complement or extend this composition would involve the danger of interrupting its unquestionable harmony. New design efforts in the proximity of the main building should, in my opinion, respect and highlight its uniqueness and create merely a reasonably neutral architectural background that complements and orders the surroundings. Thus the Manggha Museum building remains the dominant element in formal and functional terms, and it is to it that the scale and composition of the new building are subordinated. Its shell has therefore been removed as far as possible from the approach to the Manggha's main entrance and situated so as to prevent it from blocking the view of the existing building from ul. Konopnickiej. The height has been aligned to the scale of the undulating roof of the Manggha. A separate entrance zone has been designed for the Gallery, with its own stairs, a disabled ramp, and a terrace which can be used for exhibition and artistic activities outside the building.

The situation of the terrace augments the public space in front of the Manggha and creates an additional urban interior, delimited by the façades of the two buildings. The space under the terrace accommodates an indoor car park. The parapet line of the first floor is a continuation of the edge of the Manggha Museum's northern wall, and the height of the roof – the parapet of the upper storey – is below the level of the upper part of the skylights in the Museum building. Made up of simple glass and sandstone surfaces, the façades are composed in correspondence with the analogous forms of the existing outdoor architectural elements in front of the Manggha – the ramps and the stairs.

IV House / MESURA

Architects: MESURA
Location: Calle Tamarindos, 3, 03296 Matola, Alicante, Spain
Contractor: Construcciones y rehabilitaciones Mettas
Project Year: 2015
Photographs: Pedro Pegenaute

IV House, designed by MESURA is an ambitious work marked by its long period of design and building, designed from the location&demand and formalized from its systems&details.

It is located in the countryside around the city of Elche, in a hot and humid climate and barren landscape. The site consists of an existing home in the center of the field, leaving undefined surroundings: bad quality spaces and no exterior-interior relationship. The first intention will be to limit the exterior spaces through a new built volume generating a triangulation between this, the existing house and the pool. The intention is to separate the lived spaces of intimate spaces in relation to the climate, solar orientation, visual, topography and vegetation.

The ingredients introduced into the architectural equation (adding to implementation intentions in place and the response to the program) to IV are: structure, human scale, modulation and the atmosphere. All these ingredients are embodied through two main systems in place: Walls + Vaults and service packs.

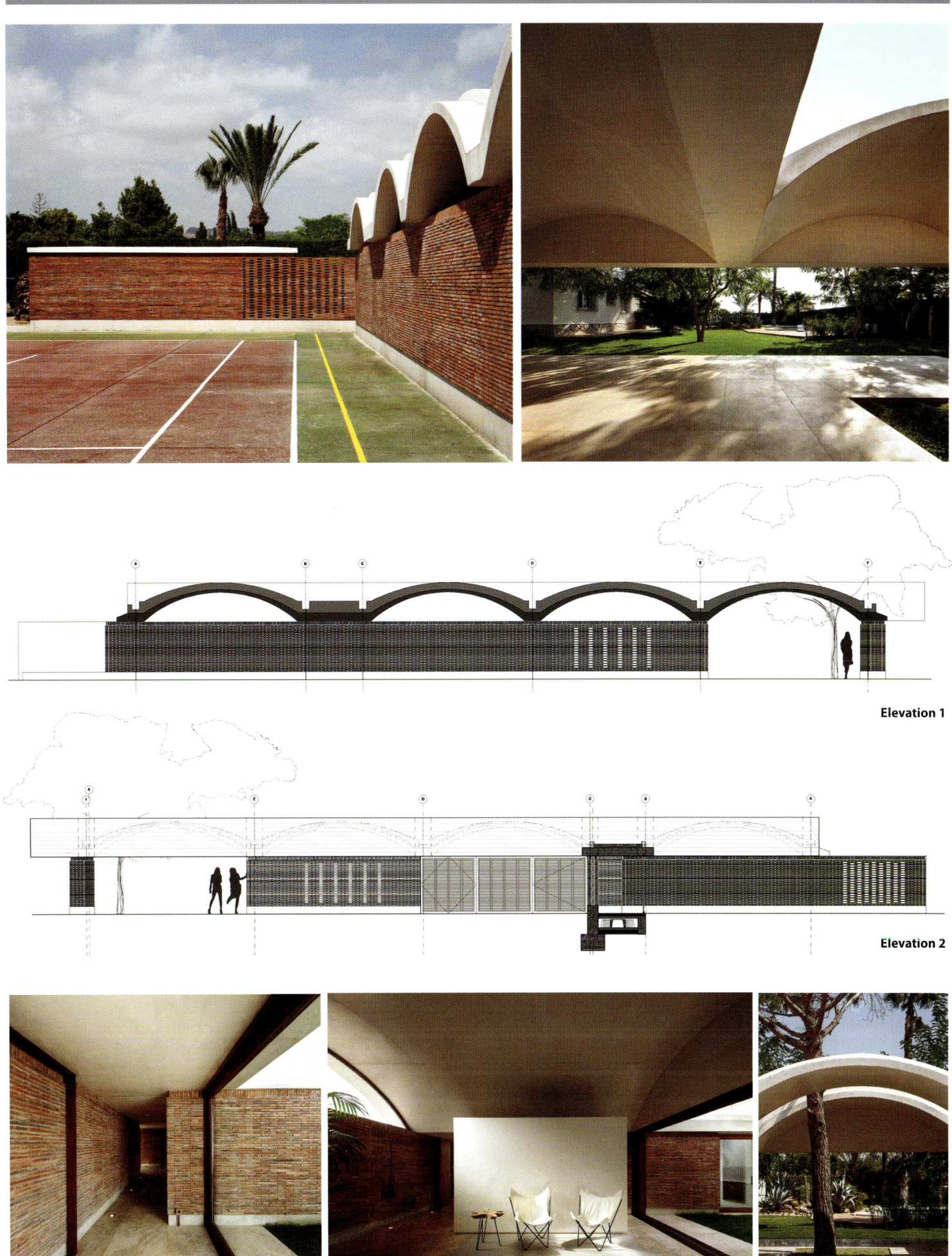

Elevation 1

Elevation 2

The ultimate ambition of the work is to be able to master every centimeter of the final result. The details that have been achieved after hours planning and researching the industry, getting define mechanisms, installations views, respect for the existing tree, (both internal and external) carpentry, hardware…

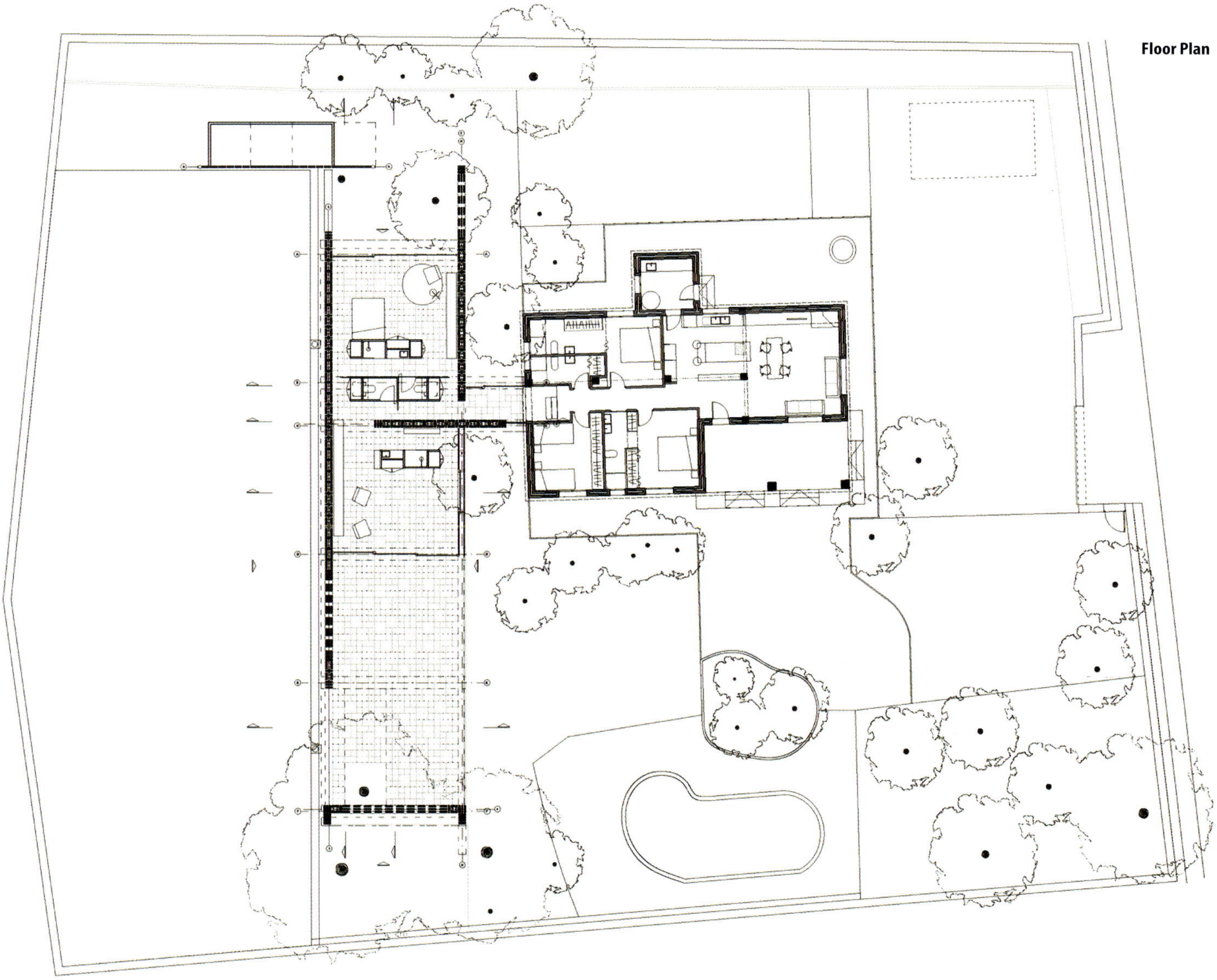

Floor Plan

Library for Qujing Culture Center

Architects: Hordor Design Group, Atelier Alter
Location: Qujing, Yunnan, China
Area: 18800.0 sqm   **Project Year:** 2015
Photographs: Courtesy of Atelier Alter
Project Architects: Haipeng Guo, Langtian Weng
Design Team: Haipeng Guo, Langtian Weng, Zhenqing Que, Ling Zeng
Architect in Charge: Yan Huang
Design Architects: Xiaojun Bu, Yingfan Zhang
Cost: $13,419,997

According to Aldo Rossi, every city needs a study room. A library is the study room for a city. Michelangelo's Laurentian library brings up the question: is the library the interior or exterior of a city?

Woman and man, rich or poor are all welcome to the library. It is a socially inclusive heterotopias space in modern cities. In a way, a "library" is the "house" for the collectives. Library is a place for thoughts. As it concerns with the classification of knowledge, library is also a critical program to a civilization. We attempt to reinvent the program of library by breaking down the separation between reading and storage areas. By having the two spaces intertwined, an abstract field of knowledge is formed, one could easily move from the perception space to the projection space.

There is a direct connection between the pattern of circulation and the trajectory of one's thought about knowledge. The act of circulation becomes an act of creation. Cross-connections between different knowledge are spontaneous, and inter-disciplinary studies are encouraged. Critical thoughts are latent in this matrix of knowledge. We base our formal transformation on the architectonic of a "vertical folded-plate." As we designate the solids to be the structure and storage, the voids in-between are collective space for reading and gatherings.

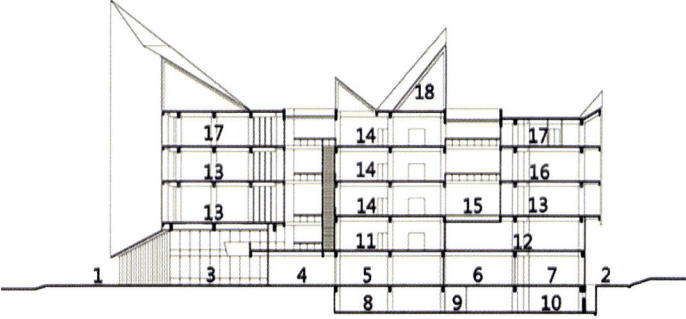

1-1' Section
1. Library entrance
2. Business office entrance
3. Entrance area
4. Foyer
5. Exhibition area
6. Collect and edit room
7. Variable power distribution room
8. Logistic support room
9. Corridor
10. Book storage room
11. Electronic retrieval room
12. Visually impaired reading room
13. Readding room(storage and read)
14. Public service area
15. Cortile
16. Specialized reading room
17. Business office room
18. Equipment

2-2′ Section

1. Cortile
2. Reader foyer
3. Public service area
4. Library foyer
5. Foyer

Model 1

Model 2

Plenus Brain Center

Architects: Naoya Matsumoto Design
Location: Japan, 〒105-0014 Tōkyō-to, Minato-ku, Shiba, 3 Chome−3−10 タツノ第3ビル
Area: 1200.0 sqm   **Project Year:** 2015
Photographs: Takeshi Asano
Graphic Design: ANTTENA CO.,LTD   **Sign Design:** Takayasu Miki

This is an office where the branding creative team of Plenus, Inc. works. Plenus, Inc. manages "Hotto Motto" and "Yayoi-ken" which have headquarters in Fukuoka.

This office is in Minato-ku, Tokyo where there are a lot of tall buildings. By making it clear to have different functions each floor, I brought in designs appears from the functions as spatial representation. When you use the elevator and get out of it, each floor has different environment, I hope it stimulates the users' brain and make them feel elation to work.

I made the façade and the first floor by brick material, which has been known as a warm atmosphere characteristic since long time ago. The permanence of brick materials resonates with the historical background of this company continued from Meiji Period. I tried to express the historical strength and the permanent strength for the future of this company. I hope those people who visit to feel and see the presence of the company continues from the past and the future by different point of view.

Double House

Architects: Bokarev Architects
Location: Tyumen, Tyumen Oblast, Russia
Architect in Charge: Ilya Bokarev, Emilia Bokareva
House Area: 97 + 97 sq.m.　　**Project Year:** 2015
Photographs: Maxim Denisenko, Artem Rich, BIlya Bokarev

A small plot of building land (600 square meters) is situated in a district of private houses in the city of Tyumen. The task for the architects was to design a two-section house for two brothers. The authors have fulfilled the basic requirements to design a separate entrance and provide car accessto each section, have a small terrace on the south side of the house and a lawn in the backyard that could be shared by the owners, and include a parking lot for two cars along each section. The house is designed to accommodate three or four people in each block 97 square meters in area. There are day stay rooms on the ground floor: a living room, a kitchen, an office, a bathroom, a boiler room, and a storage room under the stairs. The second floor holds two bedrooms and a bathroom.

On the side facades decorated with dark gray profiled sheets there are no windows due to the close proximity of adjacent buildings. The front and rear facades are covered with pale wood and white composite panels. Rusty metal is used in the decoration of the staircase. The architects extensively used local building materials.

South facade
0 1 5

Eastern facade
0 1 5

Isometry
0 1 5

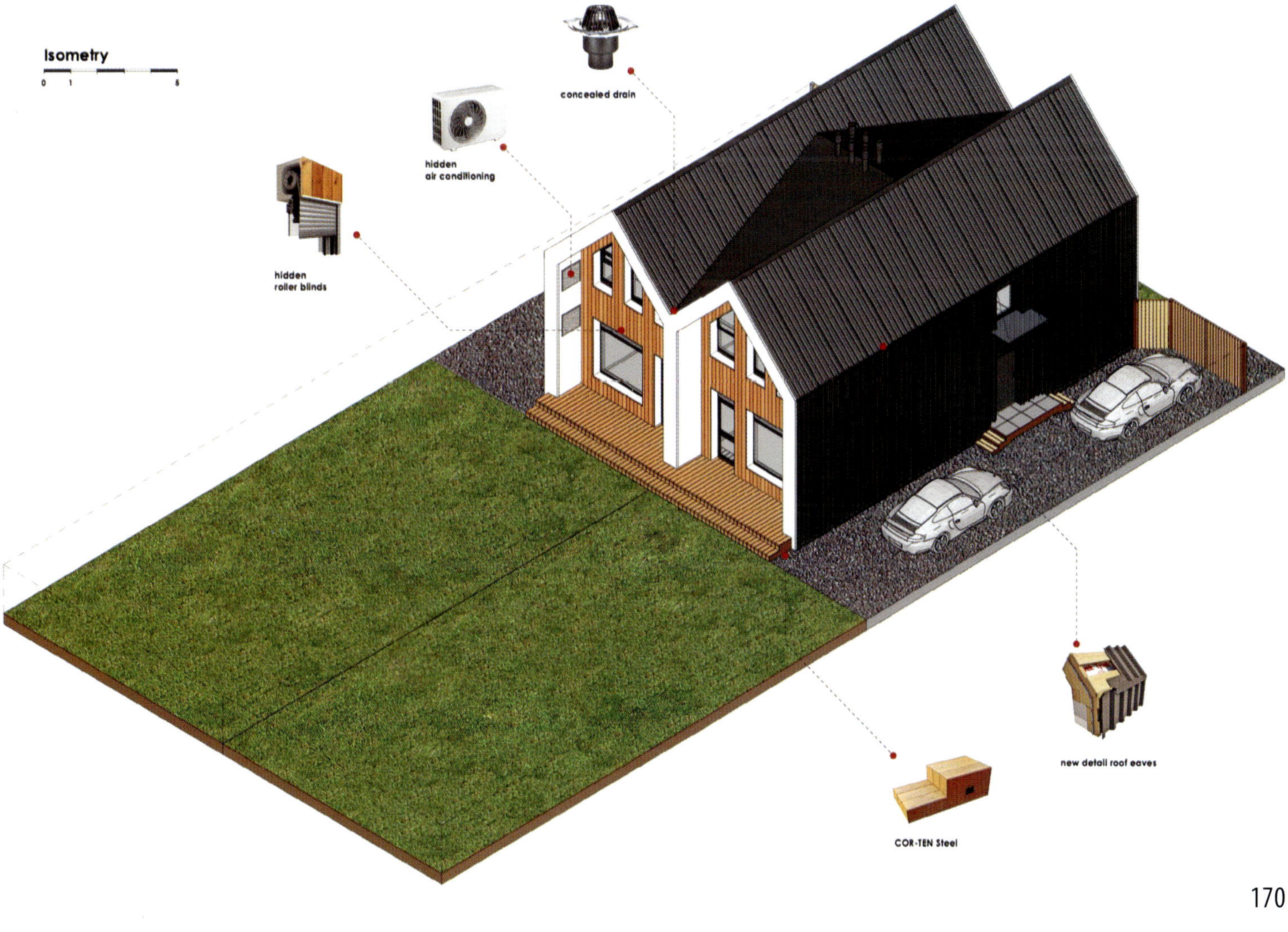

- concealed drain
- hidden air conditioning
- hidden roller blinds
- new detail roof eaves
- COR-TEN Steel

All external engineering equipment including automatic shutter boxes is concealed under the wooden facade to keep the appearance laconic. The ledges of walls have niches for outdoor units of air conditioners. The downpipe is also hidden inside the wall. The roof to wall attachment was specially designed for this project. The architects initially made the necessary drawings and a 3D model; then, a full-size model was elaborated.

Section 1-1

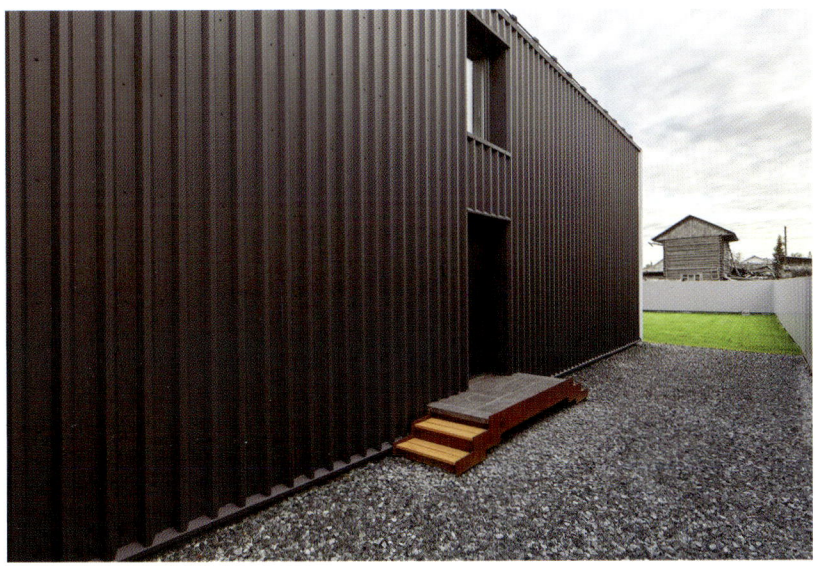

1st Floor plan

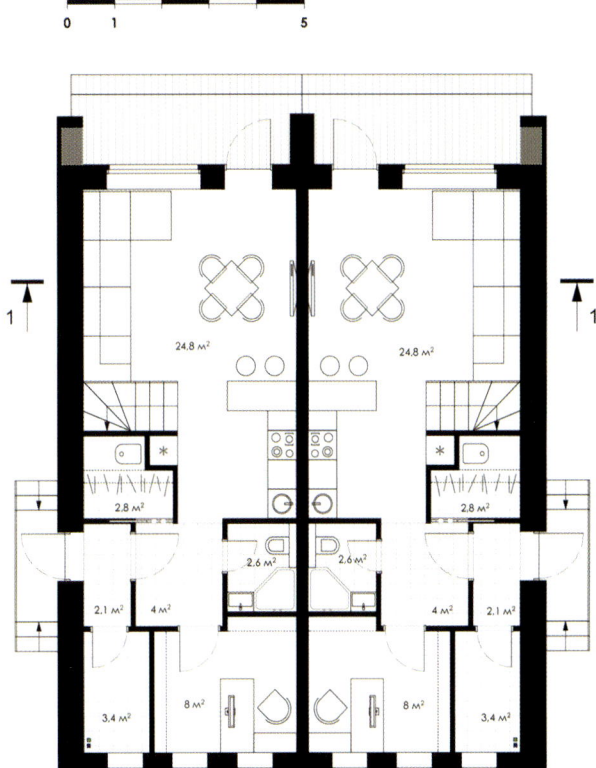

2st Floor plan

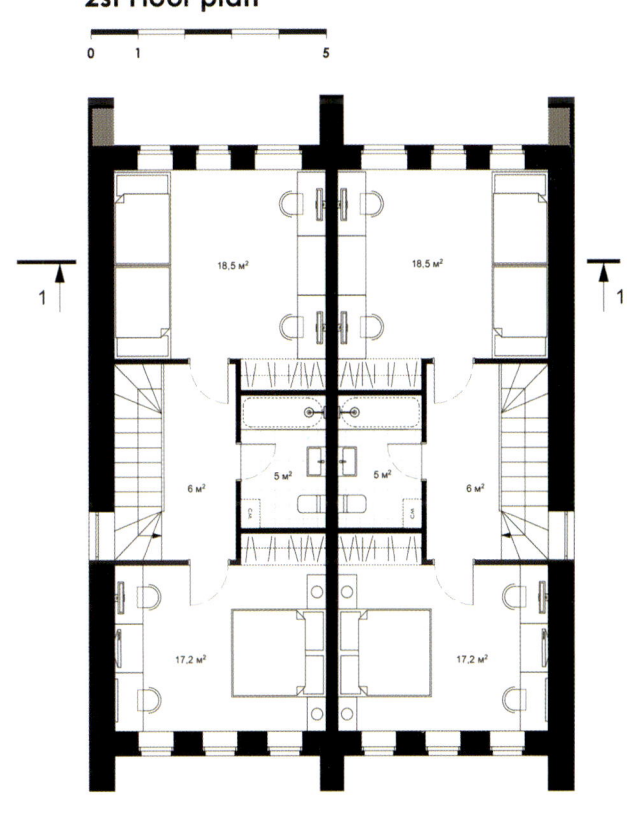

Family House In Palanga

Architects: UAB Architektu biuras
Location: Vanagupės gatvė, Palanga, Lithuania
Architect in Charge: Gintautas Natkevičius, Adomas Rimšelis, Sigita Kundrotaitė
Area: 253.0 sqm   **Project Year:** 2014
Photographs: Leonas Garbačauskas

This is a four-member family house in a seaside resort town Palanga, Lithuania. The narrow site is located approximately a kilometer of pine forest away from the sea. It features a slope and is framed - by a street on the slope foot and a forest wall on top of the hill, towards the sea. All living spaces are lifted above the street level and focused on the forest, while the utilitarian spaces are positioned on the lower level – consequently, establishing the reaction to the intricatelandscape besides sparing as much of the site for outdoor activities and for psychological perception of private spaciousness.

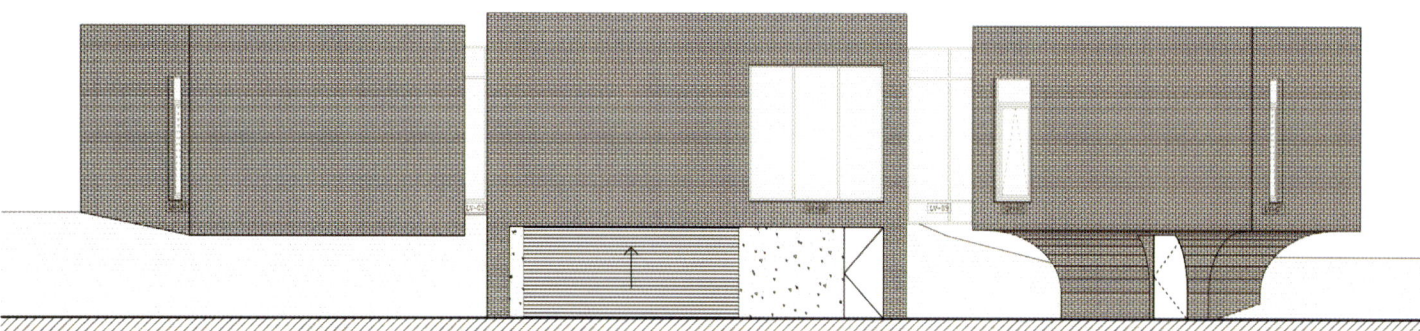

Elevation 1

Elevation 2

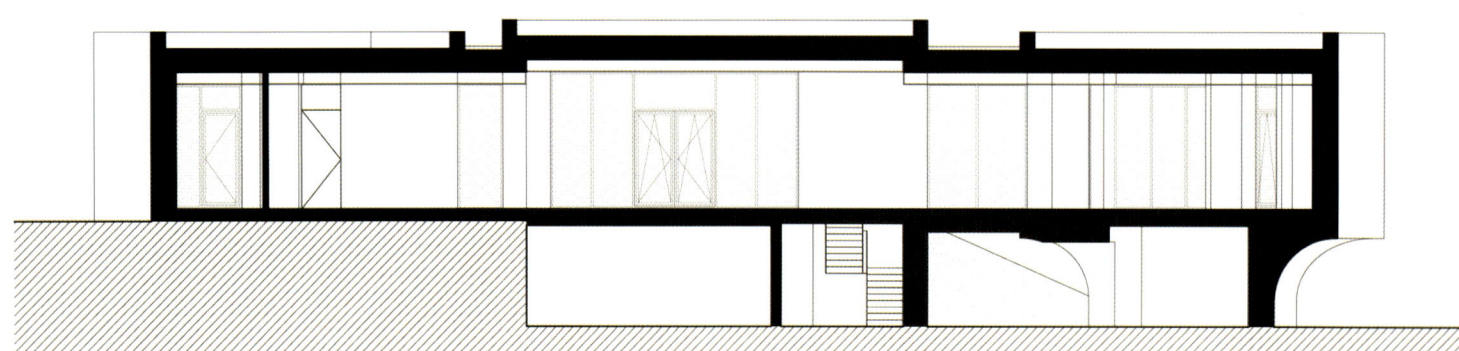

Section

The three volumes are united with a central transitional axis. Moving along it one can feel the slight changes in spaces (floor finishing that follows the direction of the axis throughout the house further enhances the sensation) and it translates into the internal scenario of the house – a gentle pulsation of spaces, light and orientation, attention to the surrounding nature, arrangement of lifestyle. As the clients have a keen eye for art, there appear places to incorporate it into the interior without it disrupting the calm flow of spaces and having it compete with the main interior attribute – the surrounding nature.

Through creative discussions it was decided to avoid the impression of a bulky hovering object that a single level living area dictates, therefore, the scheme was divided into three separate volumes corresponding with three functional zones: Children rooms with a dedicated bathroom and washroom are situated firmly on the ground, while the parents' zone – a master bedroom with ensuite facilities – is lifted commandingly on a sleek tower leg, which serves as a storage space. The central zone houses a stairway, the main living areas on the first floor and a garage, an entrance hall and technical spaces on the ground floor.

The dismantling of the scheme also allowed for delicate adjustments of orientation for all elements accordingly. The children's volume shifted to the South for light and warmth, parents' volume – to the North for a cooler climate. The two also turned away from the main volume and its terrace for privacy.

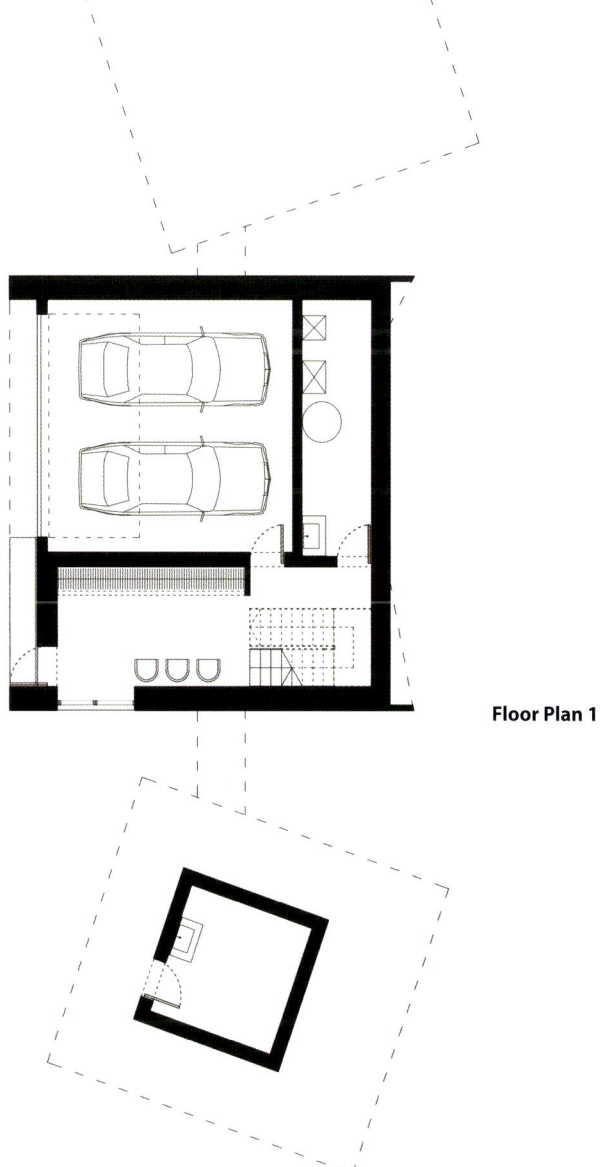

Floor Plan 1

The house becomes like fingers of a hand submerged into sand, trying to palm it; the back façade seems like a funnel – catching the light, the wind, the nature. The project strives for an image of solidity and refinement as the façade is of Belgian handmade clinker, which is stacked and glued to eliminate gaps. On the other hand, the tower leg, sometimes wittingly placed art, axis and structure all add a touch of playfulness to the house - fitting the mood of its town.

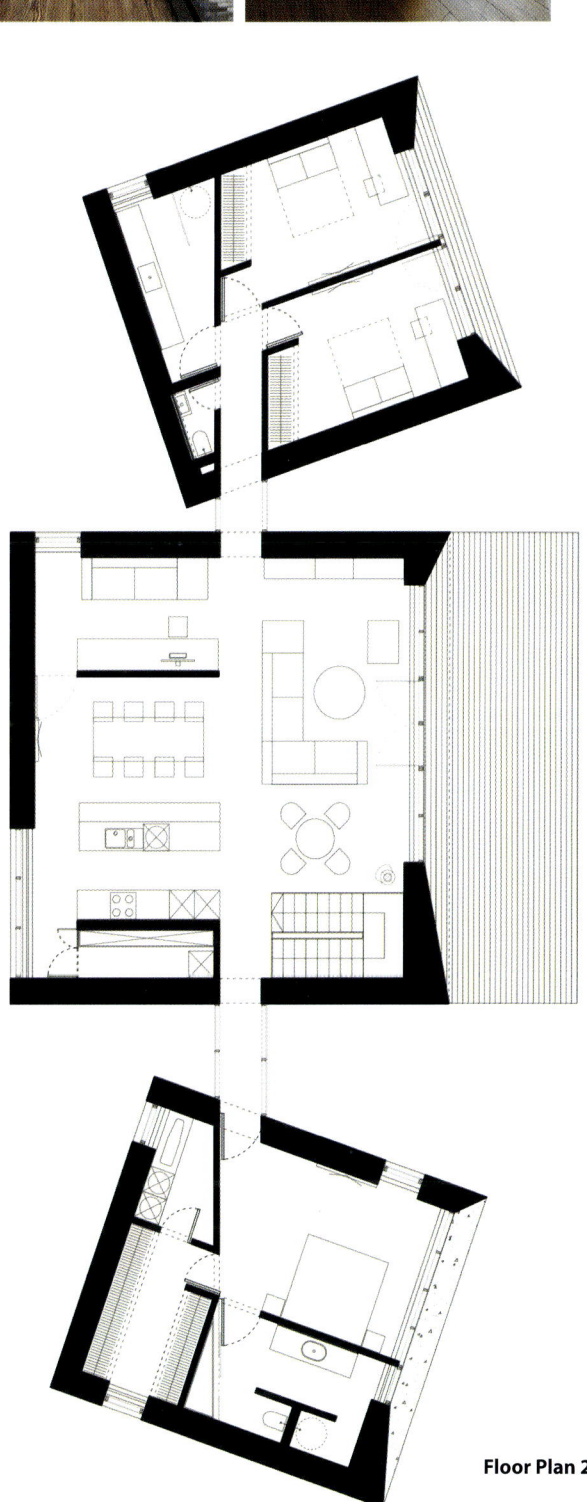

Floor Plan 2

Finlayson Street

Architects: Candalepas Associates
Location: 3/9 Finlayson Street, Lane Cove NSW 2066, Australia
Project Year: 2015
Photographs: Brett Boardman photography

The character of Lane Cove may be seen to involve an engagement with the landscape and greening of a suburb, with a subdivision pattern that resulted in individual houses on land blocks of a little less than 1000sqm. The surrounding topography of the site is one of small tree lined valleys rising to a higher ridge commercial precinct. The integration of the project into the local character has been an important consideration in the design of the project.

Elevation 1

1:250 @ A4

The project pays specific attention to a number of immediate site features such as the trees to the north and south as well as the immediacy of the neighbouring properties. The development seeks to mould a building that sits tightly within these forms to produce a sense of intimacy without a loss in amenity. The plan of the building itself is simple. It is a "U" shape with clear serrations in each wall. These serrations create privacy, sunlight access and visual interest. The strategy underlying the design was first to harmonise with the landscape and topography of the site and then optimise the amenity for the units, while creating exciting and original facades.

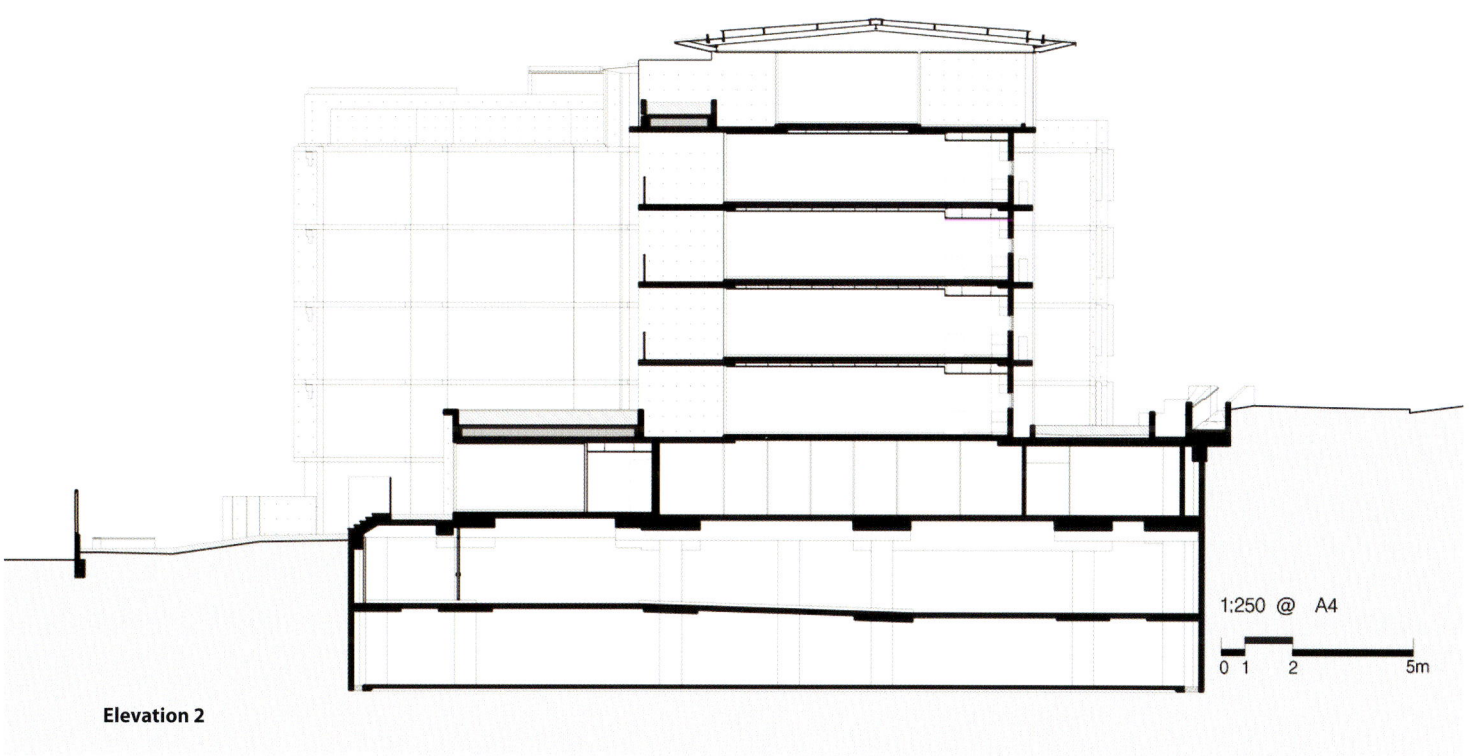

Elevation 2

Elevation 3

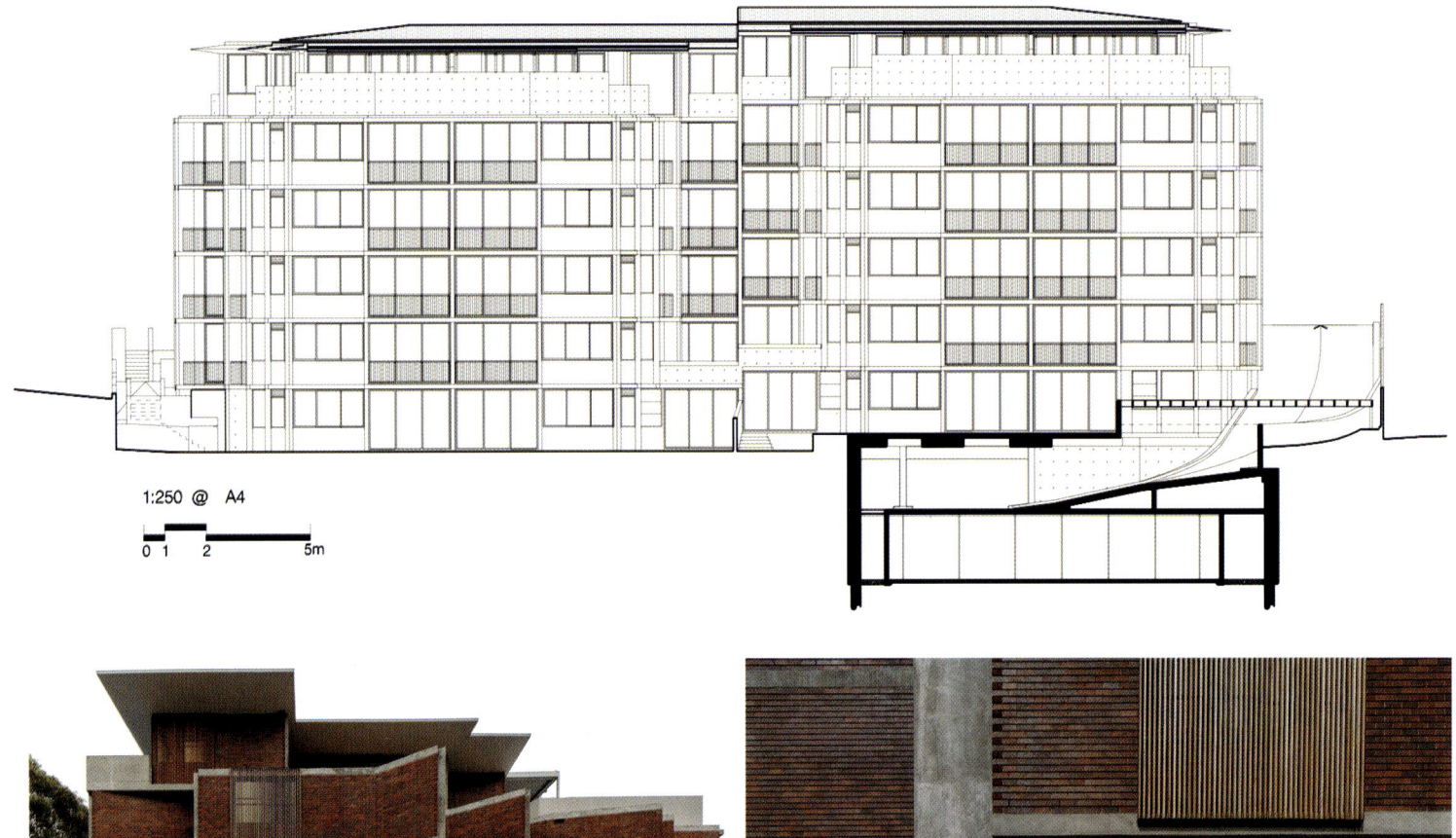

1:250 @ A4
0 1 2 5m

The project's offering to the street i.e. its south elevation, includes a series of vertical forms that are around 4.5m in width. This modulates the façade whilst also seeking to generate a robust street wall. The façade is further detailed to include natural materials such as unpainted concrete and a series of timber shutters that are purpose designed for the development. A landscape setting of shorter trees in the site at Finlayson Street acts to moderate even further, the building height at the street. The façade to the north is both more articulated through significant stepping in plan as well as affording of a larger degree of openness than that of the south. This is in part to allow for the sunlight and daylight access as well as opening the living spaces to a more private domain.

The project's offering to the street includes a series of vertical forms that modulate the façade whilst also seeking to generate a robust street-wall. These screens, in their modular and operable design, provide for both a unit-scale to the facades as well as a changing tapestry of solid and void through the pattern of habitation.

The material palette is one of both natural and robust materials with sympathy to the immediate natural surroundings. These include Australian hardwood timber, dry-pressed bricks and concrete. These materials were chosen for their innate low maintenance, robust characteristics as well as their aesthetic qualities; both immediate and as developed through the patina of age.

Espriss Café

Architects: Hooba Design Group
Location: Tehran, Tehran, Iran
Area: 28.0 sqm **Project Year:** 2014
Photographs: Parham Taghioff
Modelling & 3D illustration: Mona Razavi

Architect in Charge: Hooman Balazadeh
Project Manager: Elham Seyfi azad
Design Team: Niloofar Al-taha, Noushin Atrvash
Electrical Supervisor: Mohammad Fard, Majid Kamali
Brick laying Supervisor: Ali Akbari, Nosrat-allah Mansouri

Espriss Cafe with 28sq2 space is located in Nejatollahi street in centre of Tehran surrounded by Iranian handicrafts shops, neighboring the building of Iranian handicrafts Organization. The aim of the project was to renovate a gift shop and change it to a cafe, considering the small size of the project and its location the main idea inspired by the urban context to transform the traditional elements into an architectural interior space.

In designing the spatial diagram, the materiality concept is based on an integrated geometry continues from outside to inside. The neighboring building, Iranian handicrafts Organization with brick facade, was the inspiration to use the same material for the cafe. Concerning the small size of the project, a brick with 5*10*20 dimensions sliced into eight smaller pieces of 5*5*5 centimeters which one side of the bricks glazed in turquoise blue color. The terracotta bricks are also hygienic as they covered with antibacterial layer. Taking a glance at the existing building with two meters ceiling height on ground floor and lower height on mezzanine floor became the starting point to design the spatial diagram that could work out the issue of the existing structure.

 Full brick 5*10*20

 Dividing brick to 8 part

 Glaze on side of brick

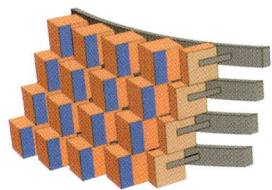 Brick work detail

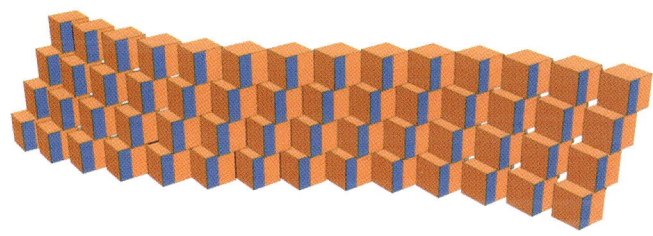

Diagram 1

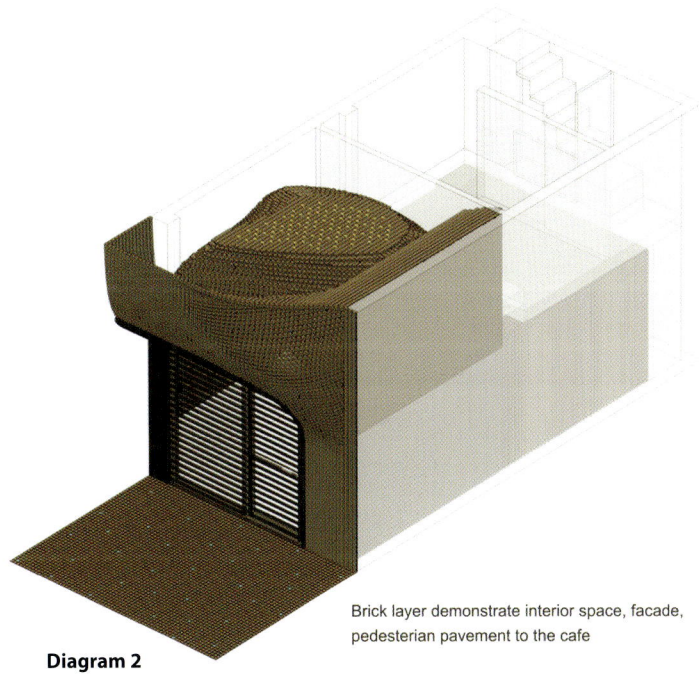

Diagram 2

Brick layer demonstrate interior space, facade, pedesterian pavement to the cafe

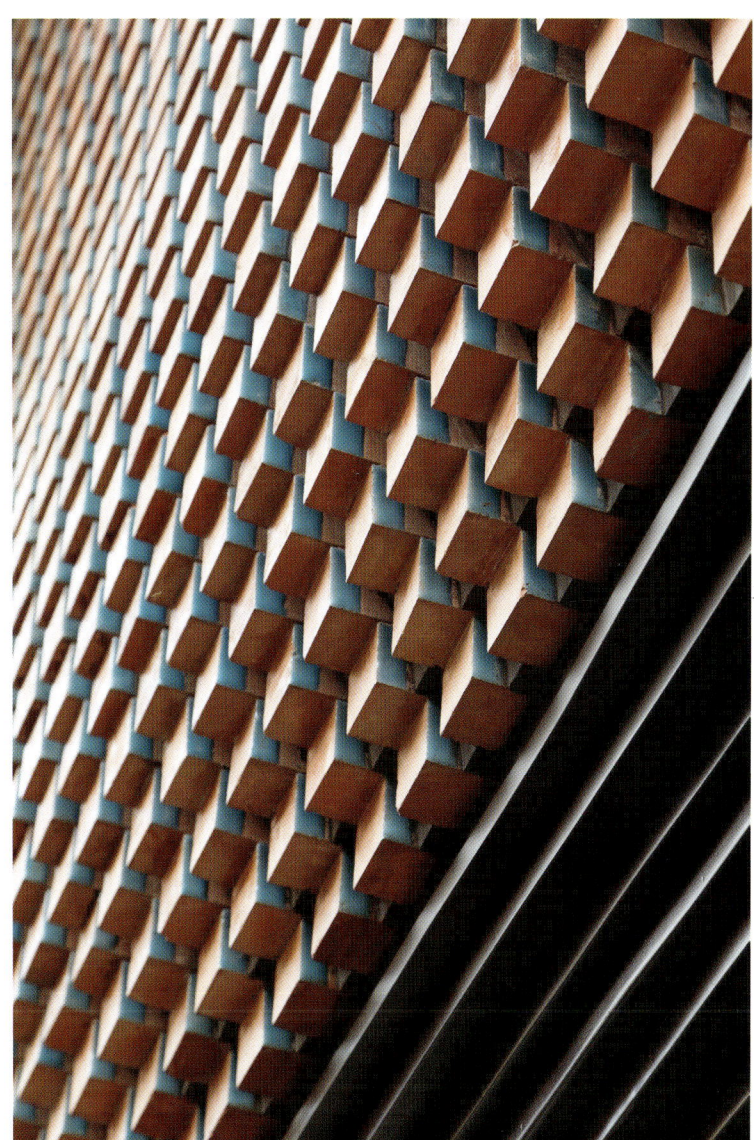

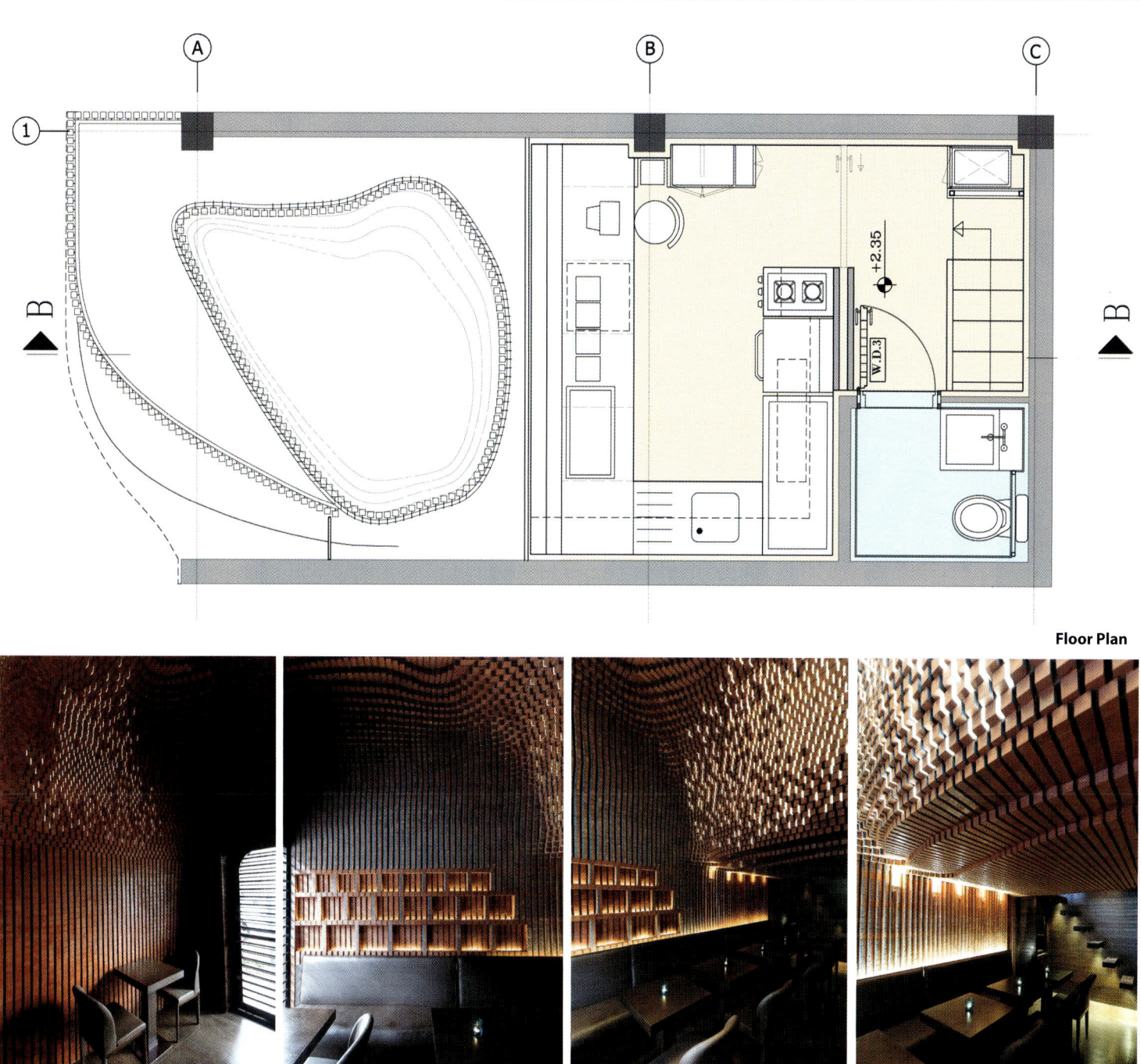

Floor Plan

At the begin half of the mezzanine floor demolished and changed into a smaller level to reach a higher ceiling for ground level, then the kitchen with services area placed in the mezzanine floor connected with the other level by stairs and a small lift. Next step was providing a ventilation system for the kitchen in the spatial diagram. The purpose was to hide the ventilation system inside the form and link it to the kitchen, so the 3D diagram stretched from outside to inside of the volume and create a connection among the kitchen, internal space and the outer geometry. The 3D diagram created an integrated structure for all the features required to work out the issues of the existing structure which also introduce morphology of brick an light as all the bricks are positioned based on the 3D diagram.

One of the main issues of design was creating a visual variation of the form in a small space. In this concept, the situation of visitors in relation to the project is significant in order to understand the form as they can differently perceive the composition of colors on the bricks regarding to their position. The turquoise blue glazed side of the bricks are facing south shaped with the monolithic geometry of brick laying that was modeled by the 3D diagram started from the pavement of the pedestrian and continues inside of the cafe. The lightening of the project is formed by the lamps provided in the gaps in between the joints of the bricks inspired by Iranian traditional architecture which translated into a modern language of design. The other material used for the interior space is dark wood includes the floor, furniture, and back side wall linked with kitchen.

Dulwich Residence

Architects: NatureHumaine
Location: 371 Avenue Dulwich, Saint-Lambert, QC J4P, Canada
Area: 2845.0 ft2
Project Year: 2015
Photographs: Adrien Williams

The clients had outgrown their 1920's house on a large lot on Montreal's south shore. They wanted a contemporary extension that would harmonize with their existing house and highlight and expose the structural brick. The existing house and the extension were separated by a glazed volume where the vertical circulation of the house is located. The existing house was re-organized to better suit the needs of the clients, where the entry and living room make up the ground floor and the children's quarters on the second floor.

Three double height spaces link the communal areas of the ground floor with the more private spaces of the second floor while maintaining a warm atmosphere in the house.

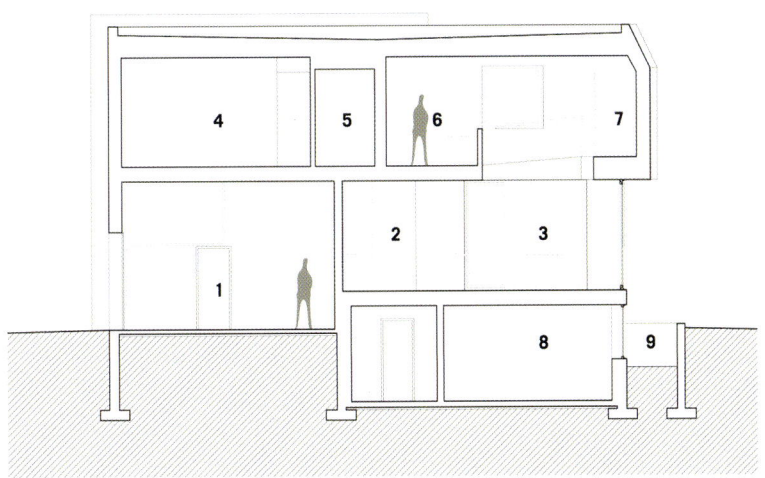

Section 1

1. garage
2. cuisine / kitchen
3. salle à manger / dining room
4. chambre des maîtres / master bedroom
5. salle de bain des maîtres / master bathroom
6. bureau / office
7. sleeping basket
8. salle de jeux / play room
9. cours anglaise

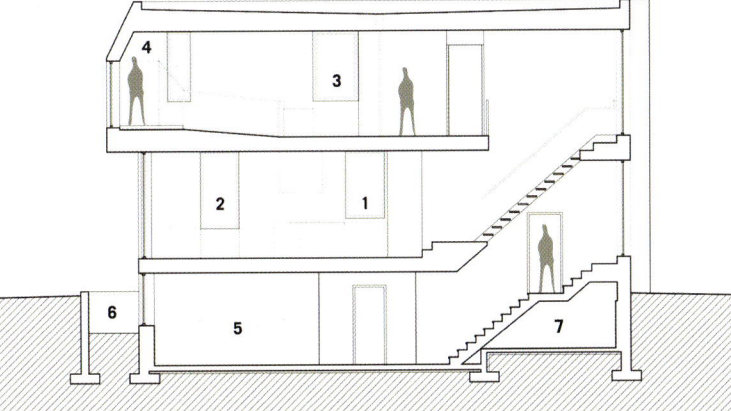

Section 2

1. cuisine / kitchen
2. salle à manger / dining room
3. bureau / office
4. sleeping basket
5. salle de jeux / play room
6. cours anglaise
7. cellier / wine cellar

188

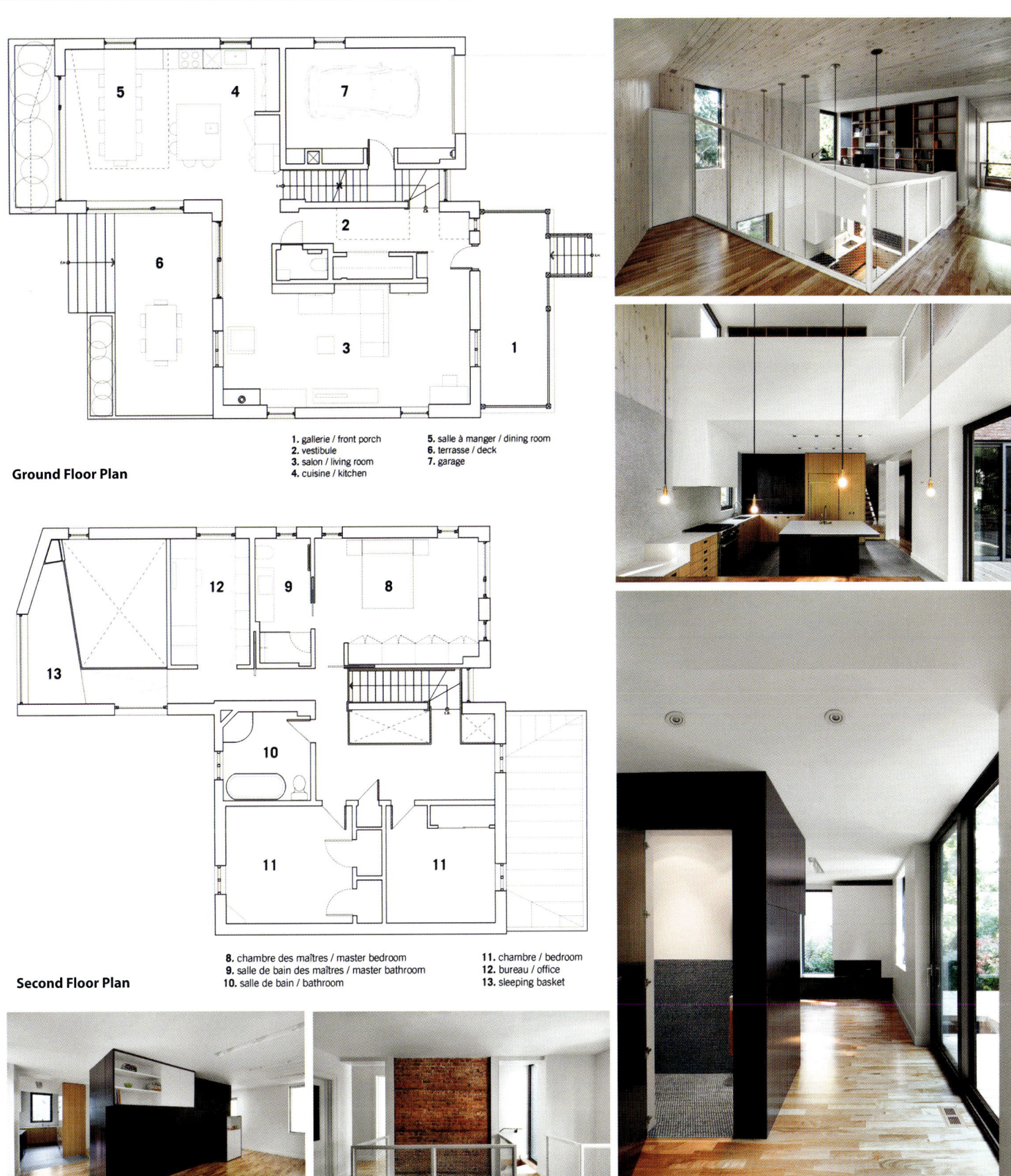

Ground Floor Plan

1. gallerie / front porch
2. vestibule
3. salon / living room
4. cuisine / kitchen
5. salle à manger / dining room
6. terrasse / deck
7. garage

Second Floor Plan

8. chambre des maîtres / master bedroom
9. salle de bain des maîtres / master bathroom
10. salle de bain / bathroom
11. chambre / bedroom
12. bureau / office
13. sleeping basket

The extension is organized into 2 intertwining volumes. A brick volume makes up the base and becomes the support for a steel clad volume that projects out into the backyard. The 'sleeping basket' is found at this projection where a large window frames the foliage. This becomes a space to relax and gaze out towards the garden as well as an area for the children to play while maintaining contact with the kitchen and dining spaces below.

Barclay School Expansion

Architects: NFOE et associés architectes
Location: 7941 Avenue Wiseman, Montréal, QC H3N, Canada
Area: 2853.0 sqm **Project Year:** 2015
Photographs: Charles Lanteigne, Maxime Pion
Project Manager (NFOE): Rafik Sidawy

Structural Engineers: SDK
Mechanical and Electrical Engineers: Pageau Morel
Contractors: Axim Construction
Artists (sculpture): Mathieu Doyon and Simon Rivest

Montreal grew rapidly during the first half of the 20th century and keeping pace was the construction of neighbourhood schools, often designed by well-known architects of the time. A decade ago, a major revamping program was launched by the Montreal School Board (CSDM) to deal with the serious need of upgrading their building stock. One of the first schools targeted for upgrade and expansion was Barclay School, designed in 1930 by Gordon and Thompson.

NFOE's architects decided to pursue a design that would be both contemporary and respectful of the historical building. They were also faced with the local residents' preoccupations about the potential loss of access to the vacant green space, just north of the existing school, which they had become accustomed to enjoying as "their park". Through quality design, the architects wished to foster a strong sense of belonging amongst the mostly immigrant families of the neighbourhood. Barclay School was awarded one of Montreal's most prestigious awards, the Grand Prix du Design, in 2014.

Program

Barclay School is an elementary school located in Park Extension, a neighborhood with a fast growing youth population. The school was in great need of additional classroom space, but it also lacked an adequate sports facility. The proposed project included fourteen (14) new classrooms, a computer and robotics room, daycare space, a multi-purpose area, administrative offices and a new double height gymnasium, to be made available to the area's residents outside of school hours.

Sustainability

The project addressed a number of sustainable issues and targeted a 25% reduction in energy consumption using the NECB as a benchmark. The building's envelope was designed to improve the school's energy performance, and a thermal wheel-based heat recovery system was installed. The new wing also features a reflective white roof and a vegetal screen.

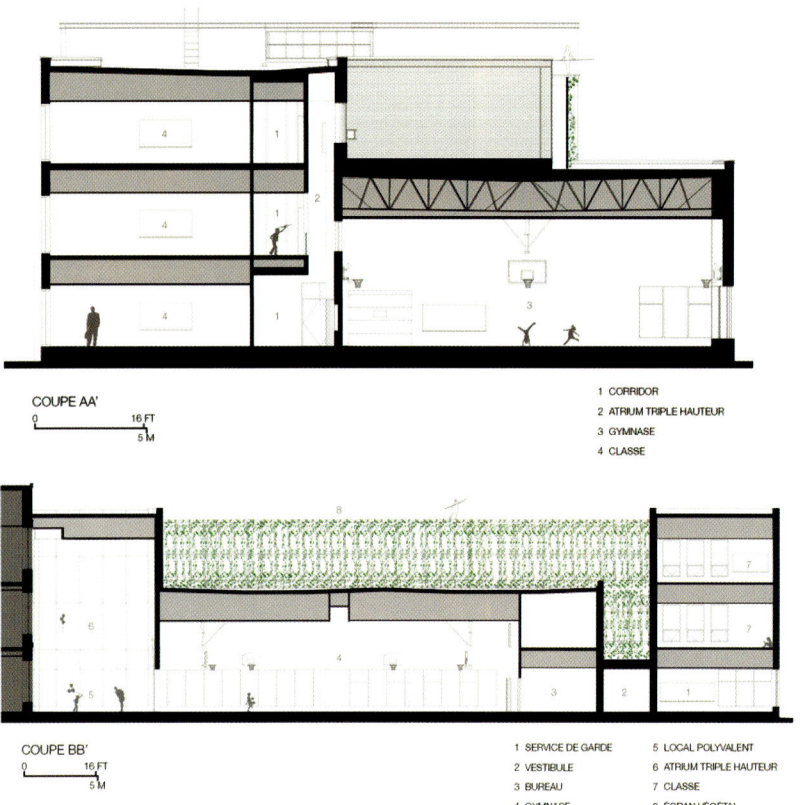

COUPE AA'
0 — 16 FT
0 — 5 M

1 CORRIDOR
2 ATRIUM TRIPLE HAUTEUR
3 GYMNASE
4 CLASSE

COUPE BB'
0 — 16 FT
0 — 5 M

1 SERVICE DE GARDE
2 VESTIBULE
3 BUREAU
4 GYMNASE
5 LOCAL POLYVALENT
6 ATRIUM TRIPLE HAUTEUR
7 CLASSE
8 ECRAN VEGETAL

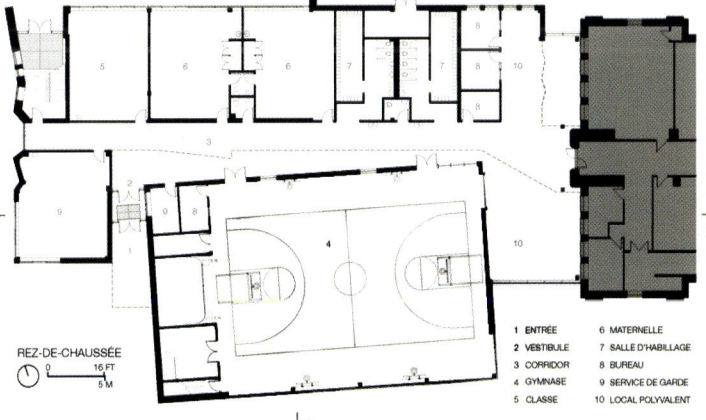

1 ENTRÉE	6 MATERNELLE
2 VESTIBULE	7 SALLE D'HABILLAGE
3 CORRIDOR	8 BUREAU
4 GYMNASE	9 SERVICE DE GARDE
5 CLASSE	10 LOCAL POLYVALENT

REZ-DE-CHAUSSÉE

Architectural Response

Due to the nature of the grounds, the most logical location for the new construction was the site immediately north of the existing building. The 2853-m2 three-storey addition was built as a lateral extension to the older building, with floors carefully aligned for almost seamless circulation between the existing and the new. A full-height glazed atrium acts as a pivot and highlights one of the older brick facades. Open to above, it serves as a welcoming multi-use space on the ground floor.

The brick chosen for the new wing recalls that of the historical building with its delightful latticework reproduced on the exterior walls of the gymnasium. The contemporary North façade appears to be partially folded, an allusion to "a butterfly unfolding its wings before flying away," to quote the designer. Windows were positioned in a random pattern, allowing the children to catch unusual glimpses of the world beyond.

The volume of the new gymnasium (two-storeys high instead of three) was built at a slight angle, altering the rhythm of the West façade and sheltering the daycare service entrance. Straddling the entire length of the gymnasium, a fine wire mesh, designed as a "green" screen, is anchored to a solid metal frame aligned with the historic building. The screen, which delimits the area available for a range of rooftop activities, also acts as a brise-soleil protecting the roof from excessive heat. Perched on top of the structure, a sculpture of a young bronze tightrope walker watches over the schoolyard.

Interiors

Communication between the new wing and the 1930 school is smooth as the architects ensured alignment of the floor levels. The ambiance is completely different however as one penetrates the atrium where natural light abounds and the eye catches subtle dashes of colour. Corridors are open to each other from floor to floor and are a source of constant surprise for curious children. A tangy green colour defines one of the gymnasium's walls and is echoed on the small storage boxes lining the corridors. It reappears on a playful emergency staircase in the back, which "might just as well have been a giant slide", according to the designer. On the second and third levels, a bright orange volume signals the children toward the rooftop above the gymnasium.

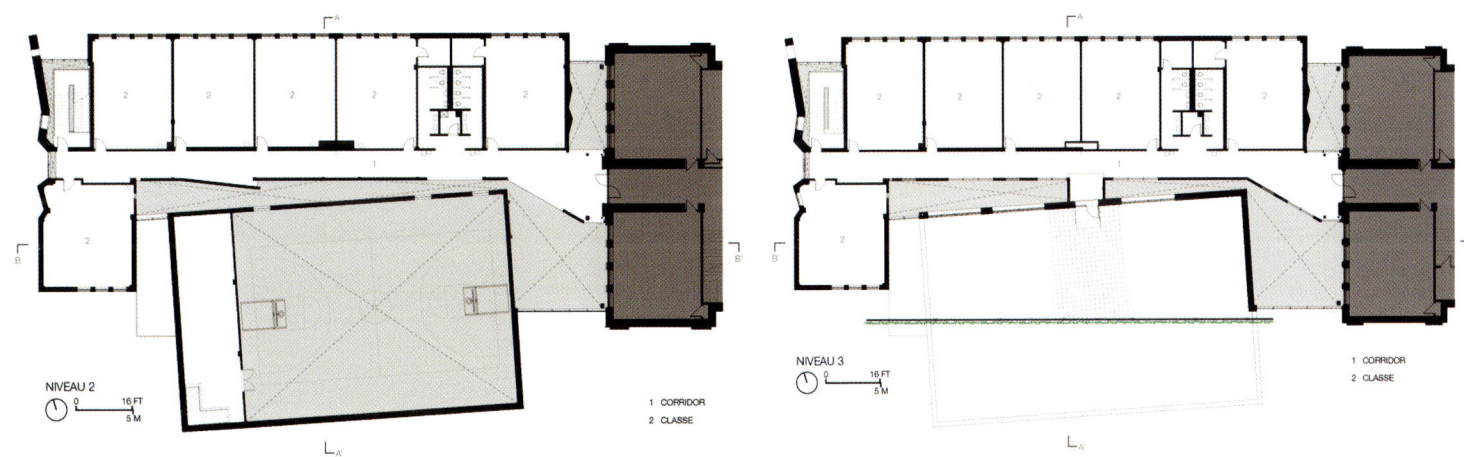

NIVEAU 2　0　16 FT / 5 M

NIVEAU 3　0　16 FT / 5 M

1 CORRIDOR
2 CLASSE

1 CORRIDOR
2 CLASSE

Bumehen-Aperture

Architects: Admun Design & Construction Studio
Location: Bumehen, Tehran, Iran
Area: 650.0 sqm **Project Year:** 2015
Photographs: Parham Taghioff
lectrical Design: Shahrouz Jafari

Design Team: Fatemeh Kargar, Mohsen Fayazbakhsh, Ramtin Haghnazar
Construction: Admun Design & Construction Studio
Project Management: Shobeir Mousavi, Amir Reza Fazel
Site Superintendence: Rashid Arbabnia, Meysam Bahrami

Located to the east of Tehran in a harsh industrial setting, Steel Form Factory's multi-purpose building is aimed at providing a pleasant space fulfilling several functions, comprising of two offices, two residential units, a showroom and several locker rooms.

When the architects were commissioned the job, the first floor structure was implemented and the team was asked to design within a tight budget, fulfilling several functions in a small land. Therefore, a cubic simple form was chosen as the dominant mass to match the industrial neighborhood fabric while reducing the expenditures.

The initial challenge was to organize basically different functions considering spaces' demands. Hence, more public functions were located on the ground floor: A five-meter height showroom on the east side with a direct access from the road, several factory's supporting spaces such as locker rooms and a praying room. Two managerial offices are located on the west side of the first and second floor with large openings to the factory hall to provide supervision. Administrative offices adjoin the CEO's office on the second floor. In order to provide privacy and take advantage of the pleasant views of the distant mountains beyond the industrial context, two residential suites are located on the third floor.

Apertures Mechanism

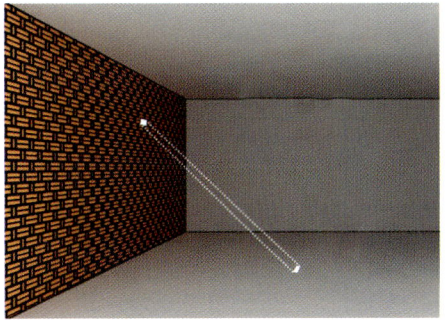
Before the apertures were covered by steel frames

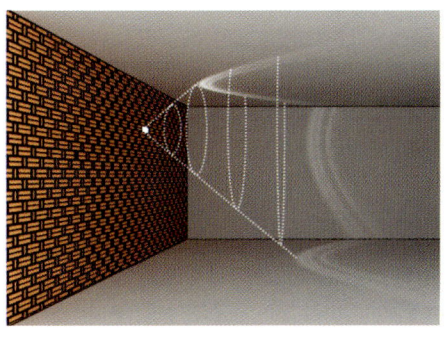
Steel characteristically diffuses light. Through the steel frames light turns into a light cone, shaping soft forms

Through this technique steel is utilized to create a peaceful environment

Process of Texture Developement
The goal was providing an innovative, low cost & easy-to-construct pattern inspired from vernacular Iraninan architecture, capable of conveying factory's identity

A Historic Building in Iran

Modules Development

Adding Steel Frames, produced by factory's waste materials

Capping frames with glass sheets

The design process initiated from answering the question of how it is possible to create peaceful spaces in such industrial fabric. The first solution to separate the interior from the chaotic exterior spaces seemed to be a mass with minimum openings which results in quiet spaces but dark and lifeless. The optimum solution seemed to be reducing the openings in size but increasing them in number that is where the idea of apertures formed.

Brick has always been used as a local material in Iran responding to environmental needs while being used in innumerable attractive forms. Using brick in this project provides the opportunity to create various textures, using merely a single material resulting in cost reduction. It can also play the role of interior finishing, leading to an integrated exterior and interior architecture. Moreover, brick adds a warm taste in contrast with the industrial cold atmosphere. Consequently, the architecture works as a landmark adding a colorful contrast to the neighborhood fabric yet being unobtrusive.

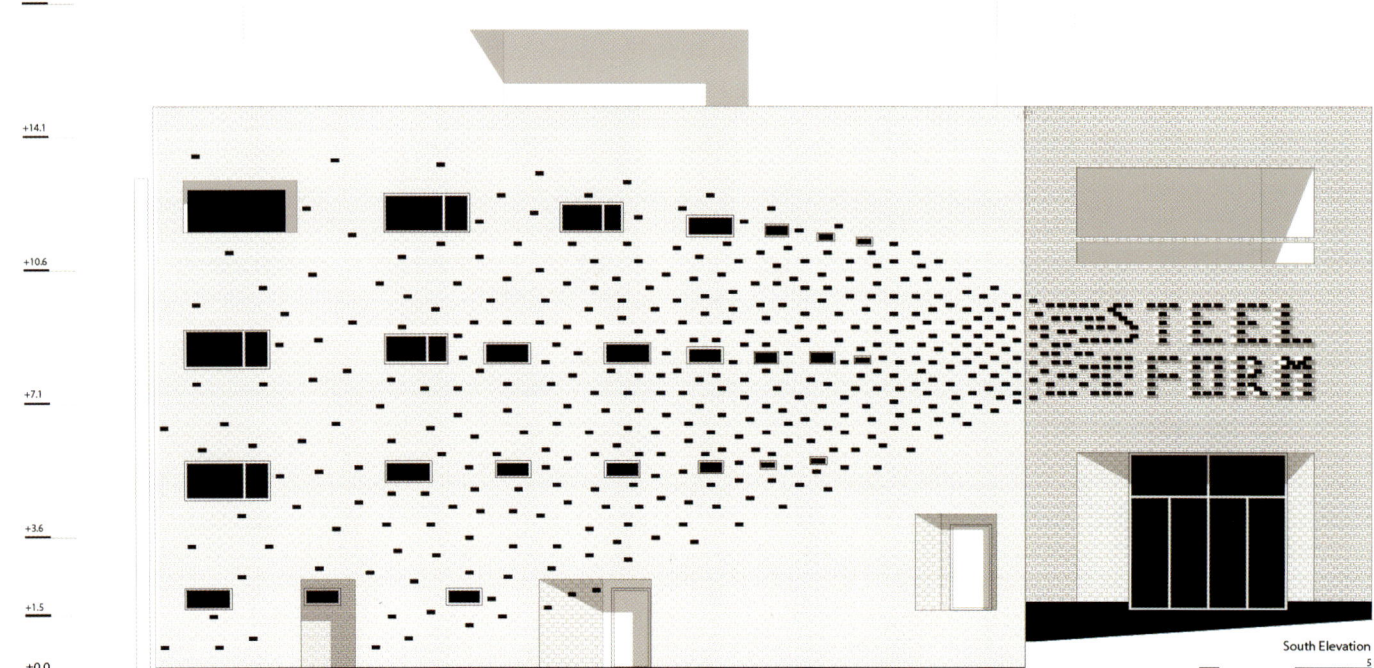

South Elevation

During the evenings when the industrial neighborhood fades away in darkness the project is still alive spreading light throughout the apertures, emphasizing on the endless motion of the Steel Form factory.

Ground Floor Plan

1 Show Room
2 Locker Room (Men)
3 Locker Room (Melting Zone Workers)
4 Praying Room
5 Factory Hall

200

120 Allen Street

Architects: Grzywinski+Pons
Location: 120 Allen St, New York, NY 10002, USA
Area: 1200.0 sqm
Project Year: 2016
Photographs: Nicholas Worley

120 Allen Street located on the Lower East Side of Manhattan and is comprised of sixteen furnished studios and four furnished apartments with a commercial space at grade. Grzywinski + Pons designed the building, the interiors, and much of the furniture. Located on an infill lot in a classically gritty LES tenement block, the property — while very narrow — is block-through and as such has frontage on both Allen and Orchard Streets. In order to accommodate the allowable floor area on such a narrow footprint, we bifurcated the mass on the north-south axis creating a slender ten-storey tower on the Allen Street side (a far wider thoroughfare with arresting skyline views) and a five-story volume on narrow Orchard Street which are contiguous in plan. While the west facing tower isn't strictly non-contextual (there are a few other far bulkier "finger buildings" on the block) we did see the upper portion of our tower as heterogeneous and decided to design the building accordingly. By treating the top five floors conceptually as a vertical augmentation to the neighborhood's historical typology we elected to create a different skin for it.

While we didn't want to pretend the "base building" wasn't brand new, we wanted the first five floors to pay homage to the existing masonry and stucco tenement neighbors. We clad the first five floors in handmade coal-fired brick and designed our own take on rusticated masonry details. Deeply framed glazing adds to the facade's texture while mitigating solar heat gain and the skin reacts to the sun's path throughout the day in a softly graphic way.

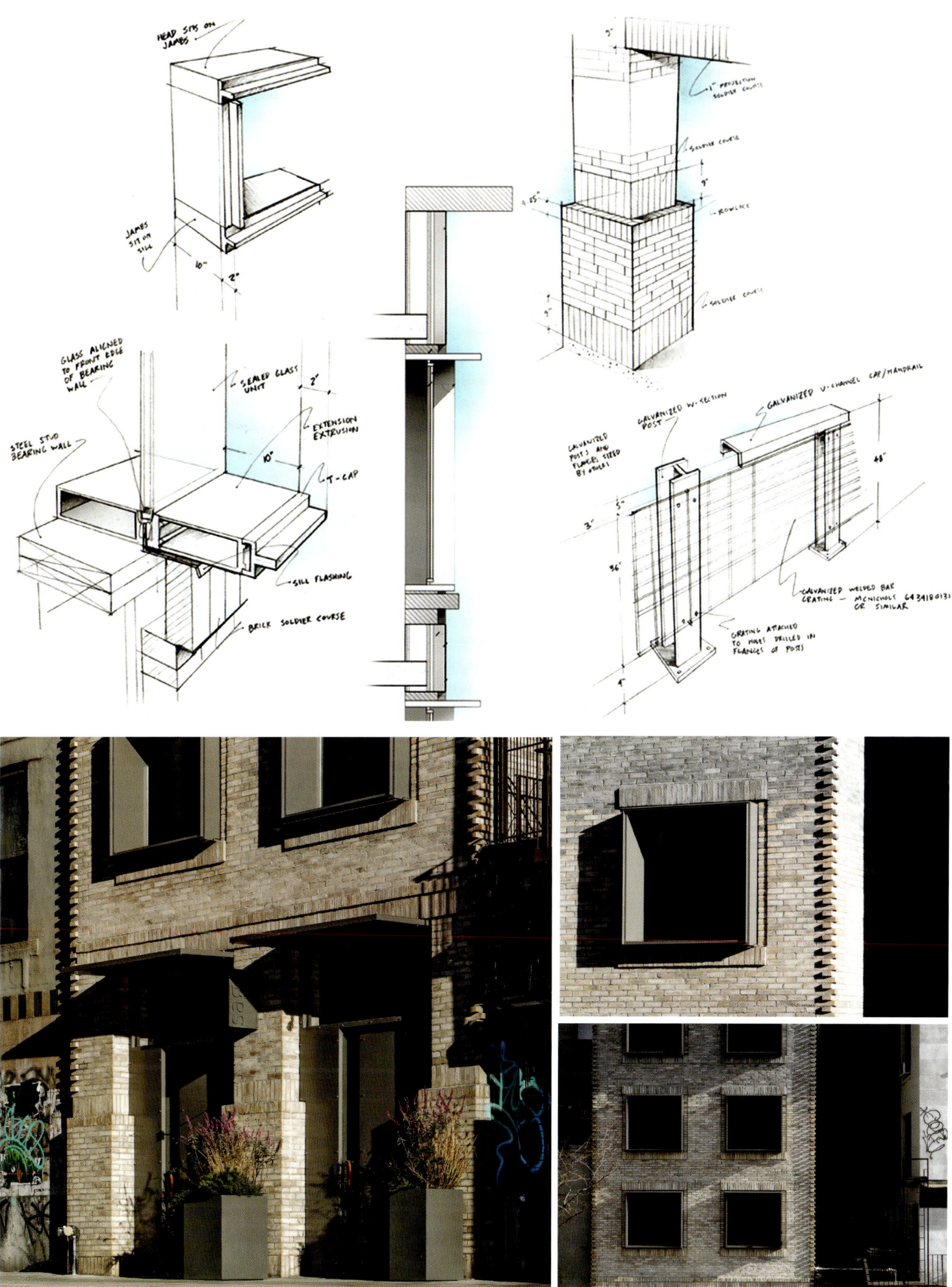

We chose bricks that subtly change hue as the building rises until they closely match the cladding of the volume above. The upper part of the tower — the volume beyond the rooftops of the adjacent buildings — is clad in zinc panels which are then surrounded by a veil of louvers. While their primary purpose is — like the window boxes below — to mitigate solar heat gain, the louvers also serve as balustrades to the balconies, create extra depth in the facade, yield kinetic shadow play and obfuscate externalized mechanical components. The colors, proportions and textures of the two envelopes reference one another and are more like siblings than strangers.

On the interior we decided to celebrate the steel structure by exposing it wherever possible which helpfully yielded more width within the constrained lot. Those girders, columns, bolts and junction plates coupled with block and masonry are balanced with a collection of warm, soft and luxurious finishes, fabrics and furnishings. We were conscious of the depth of the spaces and the special character the light has when it stretches through the interiors; grazing walls, upholstery, bedding and tile. We designed them all in such a way to extol the virtues of that long light. Cozy, comfortable, intimate rooms but with a sense of drama and a real reflection of context. These are rooms in the heart of the Lower East Side and, as pampered as one might be within them, we wouldn't want you to think you were anywhere else.

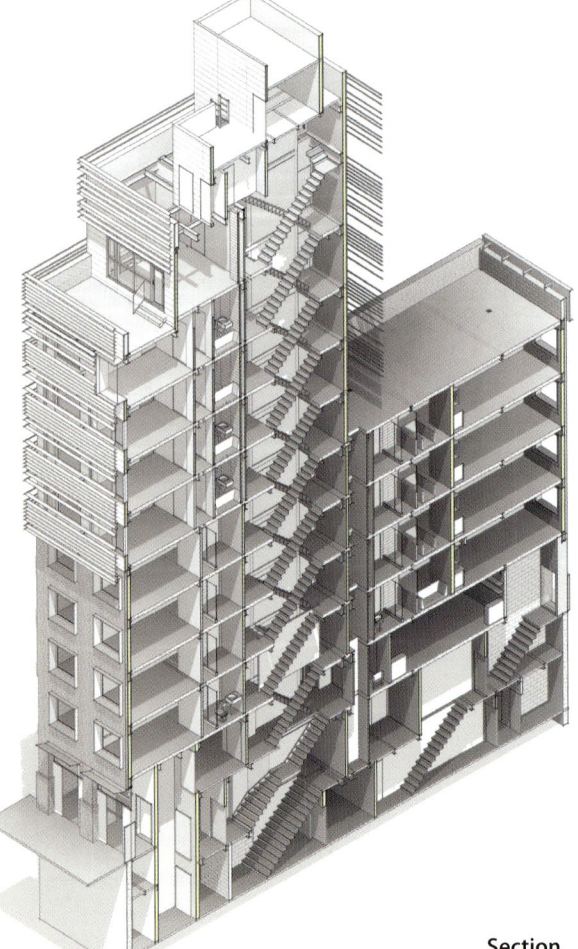

Section

BSAS Geodesy Directorate Historical Archive

Architects: SMF Arquitectos
Location: Calle 61 e/ 10 y 11 Ciudad de La Plata, Argentina
Project Architects: Enrique Speroni, Gabriel Martinez, Juan Martin Flores
Project Area: 1000.0 m2   **Project Year:** 2014
Photographs: Albano Garcia

We define a pure volume as an autonomous form that contrasts with the environment and demonstrates its representation as an institution for its originality and distinction through its own unique character that is not tied exclusively to formal and epidermal aspects but to environmental ones in the broadest sense.

The contrast between solids and voids is regulated by the layout of the openings, as these let in the necessary light only to the areas that require it. That is, where the program most needs it, balancing the functional and the formal. The block is the key that defines the buildings and their urban space. It is what determines the urban scale, its density and its uses. Within this scale and its respective grid, we start from the assimilation of the party lines where the chosen lot has a minimal public facade and a rear that is five times as large. In this context, the building stands as a single body of net lines, pure, surrounded by an environment of residential scale that will change in time.

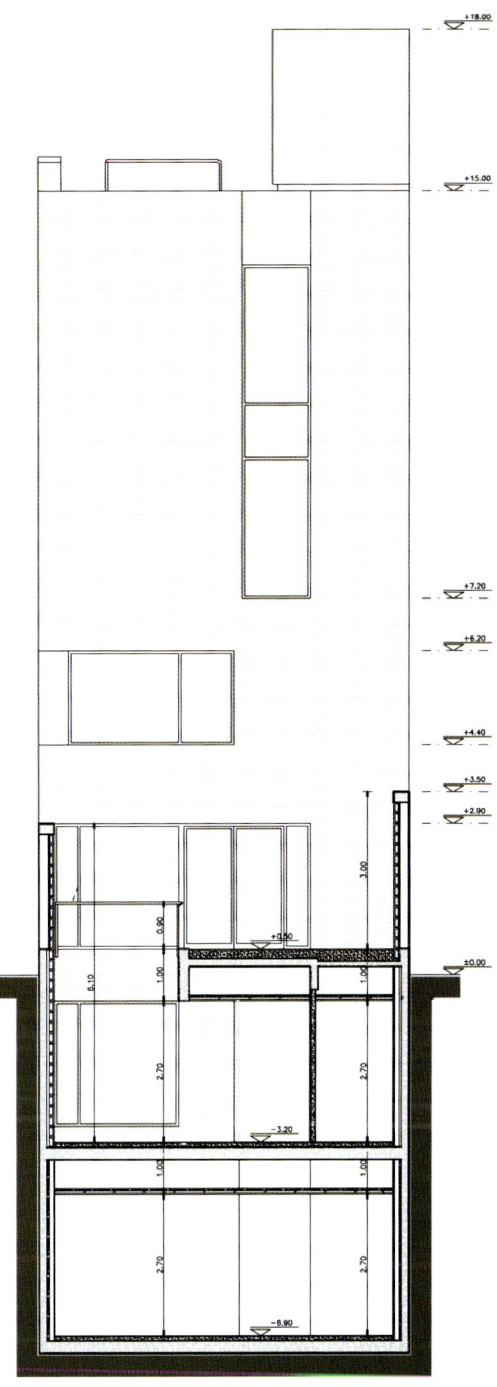

Elevation 1

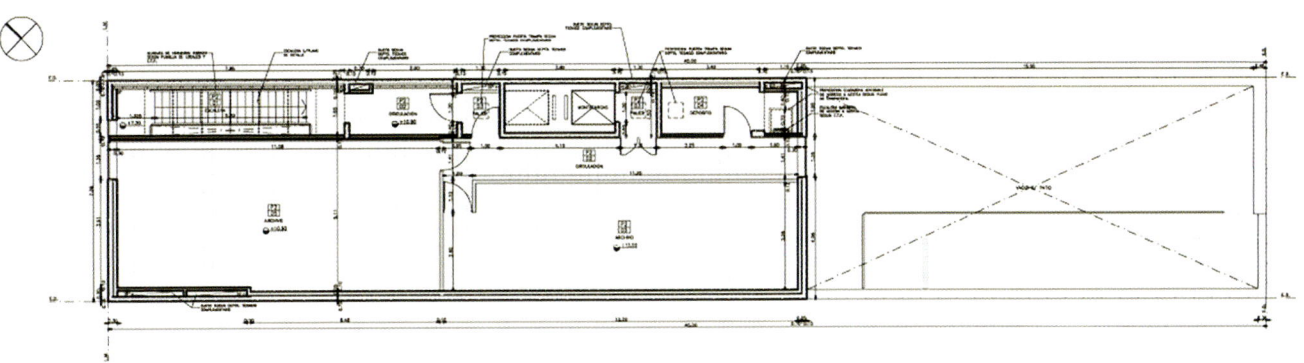

PLANTA 2° PISO
NIVEL +10.90

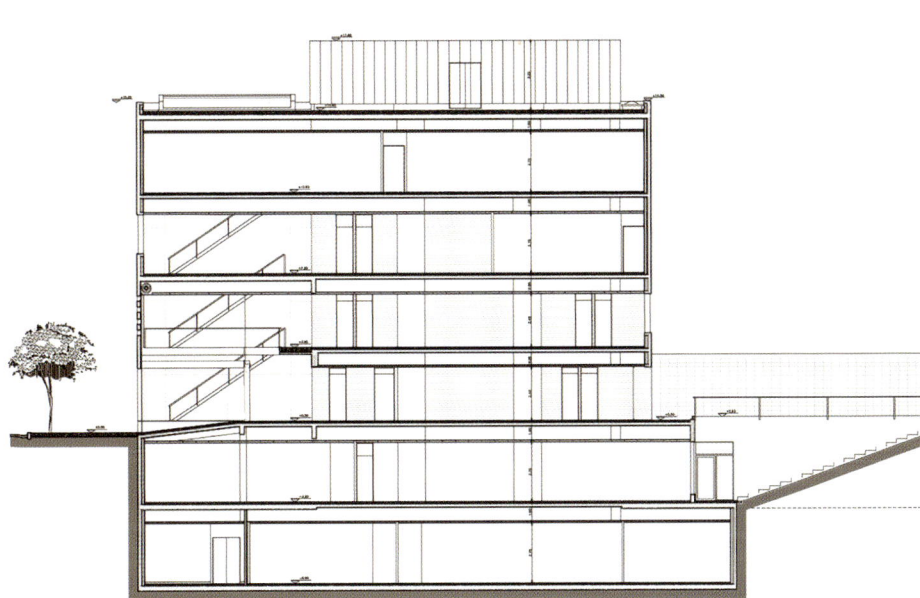

CORTE B-B

Within the proposal, the challenge is how to respond to a unique program: archives that keep cartographic documentation with a high historical and reference value, combined with activities that promote research and dissemination of that heritage. Everything must be resolved efficiently and simply.

This situation gives an outstanding physical presence that arises from its structural design and presents two faces that "communicate" with the outside in a singular way, front and back and two blind sides, the party walls. The integration in a given context is not seen as an act of mimesis but creatively. A public building should have a representative quality and its functions don't need to be expressed but instead classified.

House of Dior Seoul

Architects: Christian de Portzamparc
Location: 464 Apgujeong-ro, Gangnam-gu, Seoul, South Korea
Local Architect / Project Manager: DPJ & Partners, Architecture
Area: 4408.0 sqm   **Project Year:** 2015
Photographs: Nicolas Borel, Courtesy of Christian de Portzamparc

GFRP Façade Development: DPJ & Partners
Structural Engineer: CS Structural Engineering
Main Contractor: Kolon Global Corporation
GFRP Façade Sub-contractor: Design Base
Metal Façade & Curtain Wall Sub-contractor: Iljin Unisco

C M G W

I wanted the building to represent Dior and to reflect Christian Dior's work. So I wanted the surfaces to flow, like the couturier's soft, woven white cotton fabric. These surfaces, which soar into the sky and undulate as if in motion, crossed by a few lines, are made from long moulded fiber glass shells, fitted together with aircraft precision.

In Seoul, where the quadrangular buildings align with the avenue, and which are all occupied by leading international fashion labels, the building stands out like a large sculptural tribute to Dior, inviting everyone to step inside.

The entrance, where two shells come together, is a sort of modern lancet arch, in which two metal mesh surfaces cross in line with the clothing metaphor. Once inside, the customer makes a succession of discoveries – a feature typical of the interiors designed by Peter Marino.

Elevation 1

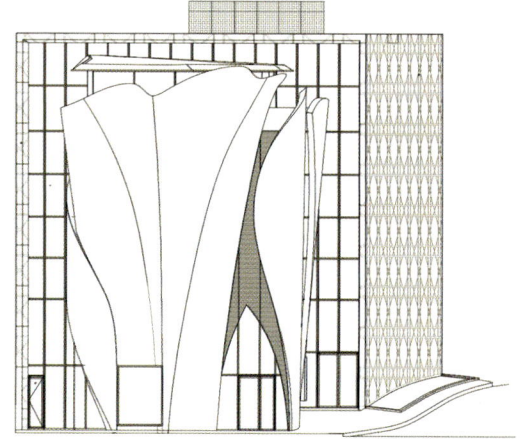

Elevation 2 *(left)*   **Elevation 3** *(right)*

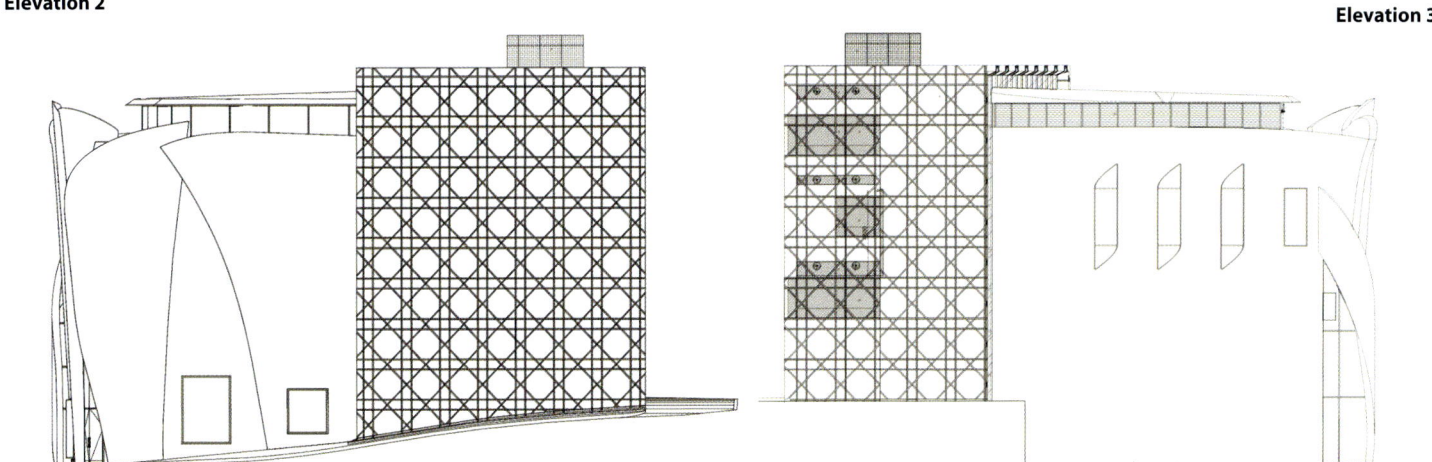

Wesfarmers Court at Curtin University

Architects: JCY Architects and Urban Designers
Location: Kent Street, Bentley WA 6102, Australia
Project Year: 2015
Photographs: Rob Ramsay

Wesfarmers Court is the first project completed within the Place Activation masterplan with the principle aim for the revitalization of the courtyard to create a flexible and adaptable space to showcase Curtin's Business School activities through promoting greater utilization of this space and as the new gateway 'hub'.

The project consists of a steel framed structure inserted between the Curtin Business School, Angazi Cafe and the Lance Twomney Theatre. The steel structure's large expanse covers the entire courtyard to provide a framework defining a functioning courtyard with perimeter perforated screens and a fully operable roof canopy that respond to the varying climatic conditions and seasonal changes. The structure has been fitted with a large media screen for events and presentations, heaters, feature lighting and a ticker tape display 'loop' displaying information and current news items.

The space has been dispersed with various loose furniture to allow for a versatile function area that can perform to a variety of events from weddings, networking events, music entertainment, outdoor lectures and presentations as well as the day to day social hub for students.

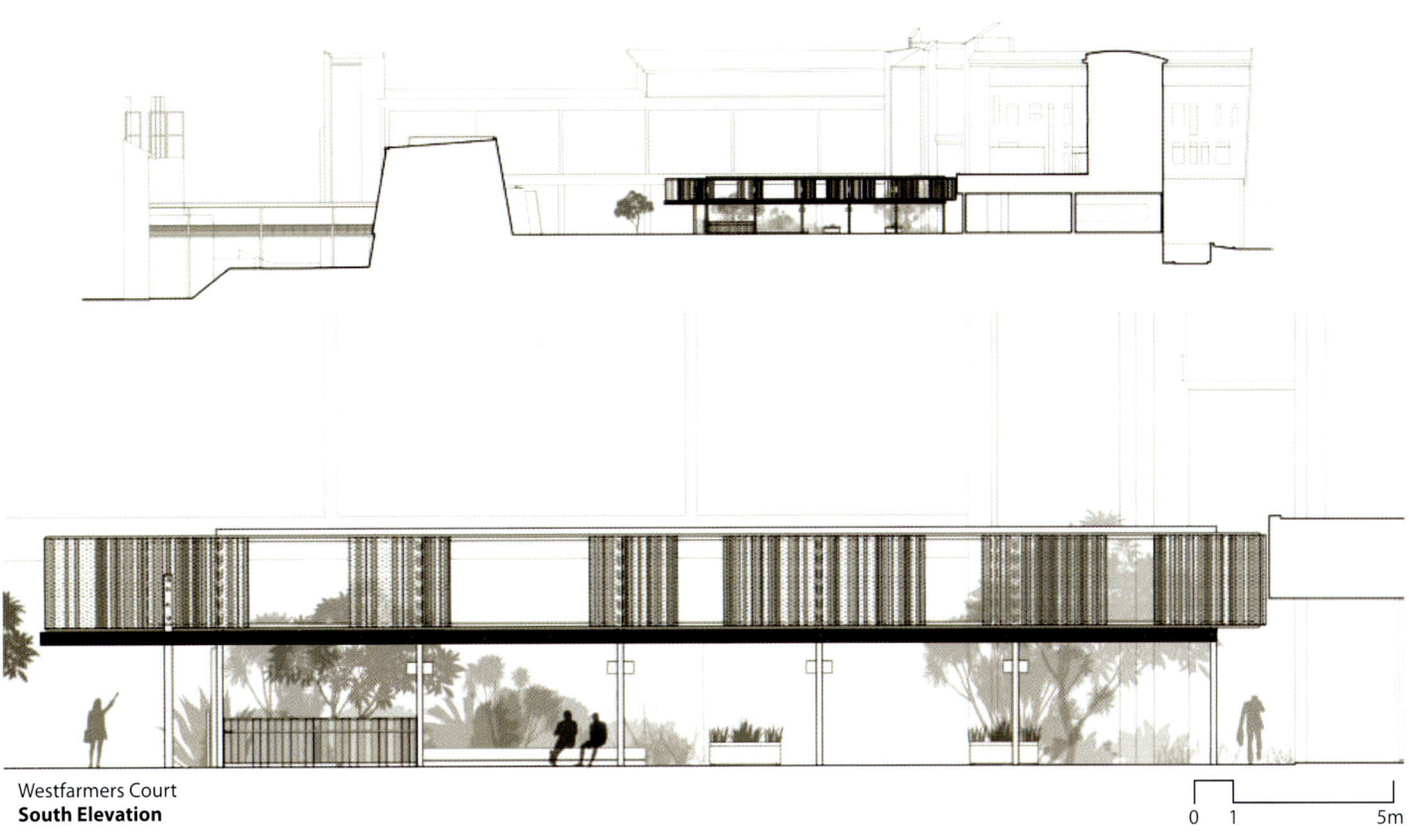

Westfarmers Court
South Elevation

0 1 5m

VAH.AX: 0.475 (+0.005 / +1.064%)

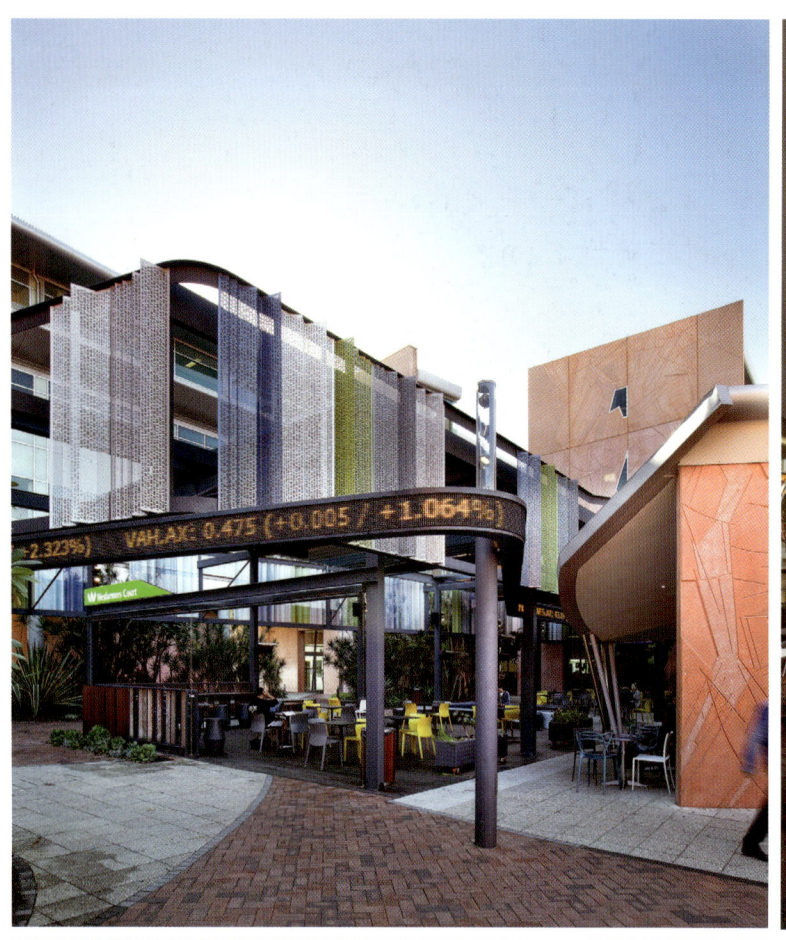

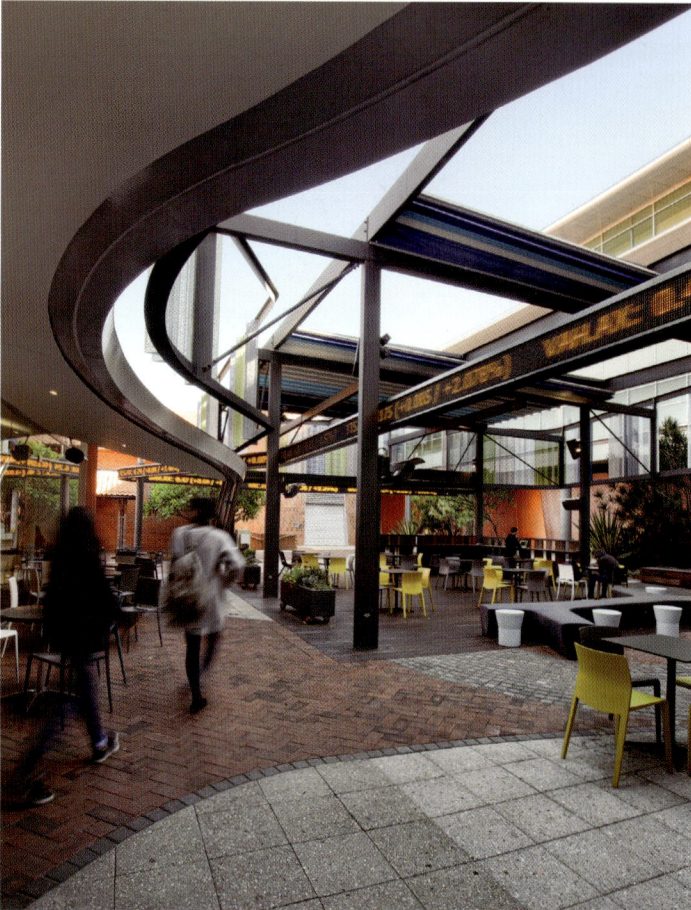

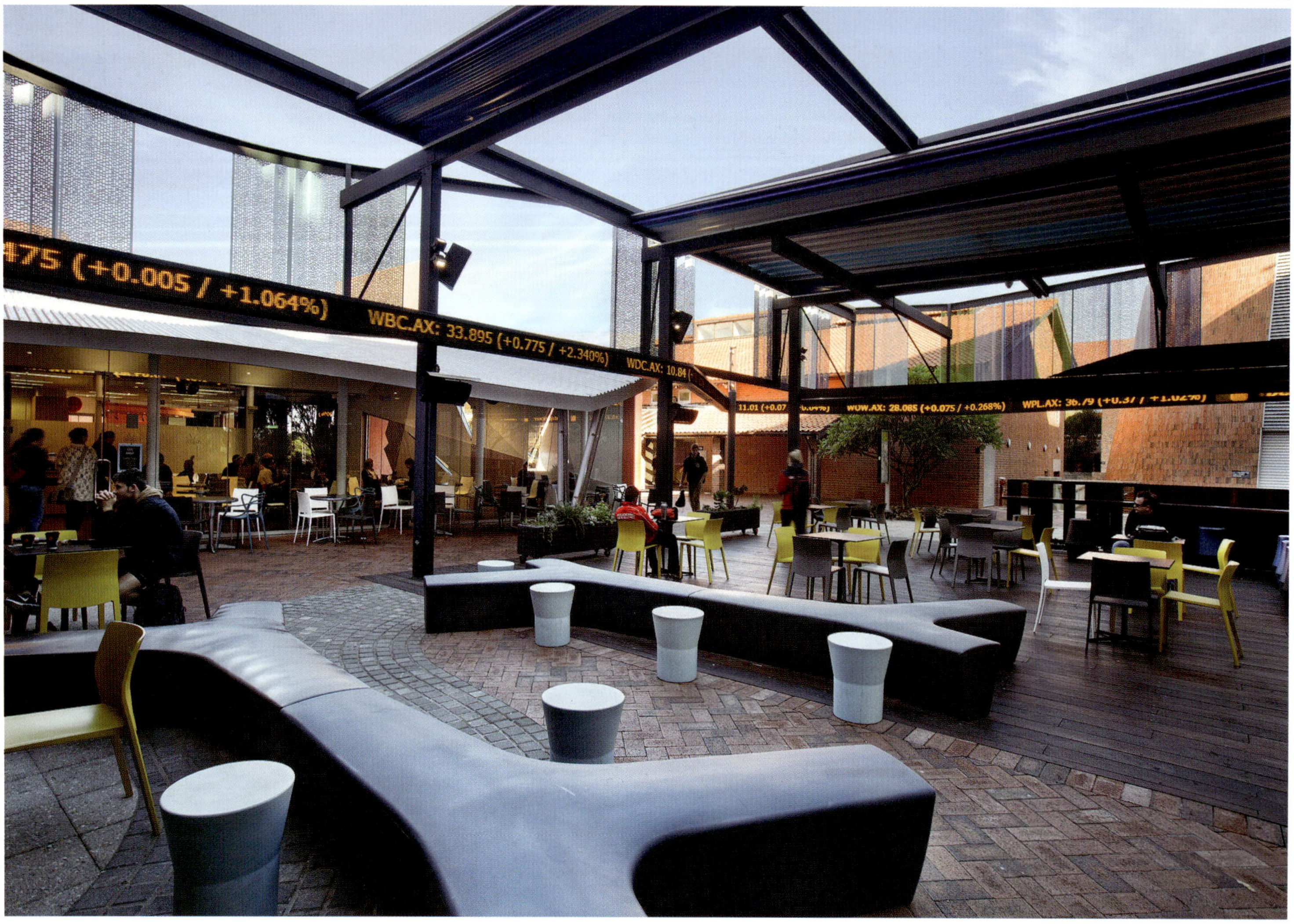
475 (+0.005 / +1.064%)
WBC.AX: 33.895 (+0.775 / +2.340%)
WOW.AX: 28.085 (+0.075 / +0.268%)

Future stages of the redevelopment of this precinct will involve the demolition of two concrete classroom 'pods' under the Business School building. This will open up the northern vista, an extension of the ticker tape display loop to form the northern gateway entry point of the 'corso' pedestrian avenue. It will also incorporate the expansion of the function areas, landscaping and various pop-up retail outlets housed within shipping containers.

The introduction of small pop-up retail outlets or a bar will serve to encourage greater campus activation and extended trading hours for the existing café. Night time activities such as movie nights or musical performances will encourage students to better engage and interact with their campus in the creation a lively, activated and attractive learning environment.

Residential and Office Building

Architects: blauraum Architekten
Location: Hoheluftchaussee, Hamburg, Germany
Team: assmann beraten+planen GmbH, Hamburg; HHP Ingenieure für Brandschutz GmbH
Area: 2123.0 sqm   **Project Year:** 2015
Photographs: Werner Huthmacher

The new building composed of rental units and retail space is located on Hoheluftchaussee, one of Hamburg's main arteries leading to the city center. The primary concept is to fill a vacant plot and create high quality living space while taking into account the high traffic volume. The building is composed by one central nucleus with a circulation of nine apartments. Each floor contains two apartments except the top floor: there one apartment with a huge roof terrace was planned. A focus lies on a southward facing open concept with a view onto the interior courtyard thus avoiding the busy street on the opposite side. Each apartment has a balcony facing south as well as a loggia facing north - the Laubenzimmer. The floor plan was created open and widely. The sleeping rooms are oriented towards the streets while living and cooing opens to the southern side to the courtyard. This first floor was concepted as an open office unit with a roof top terrace.

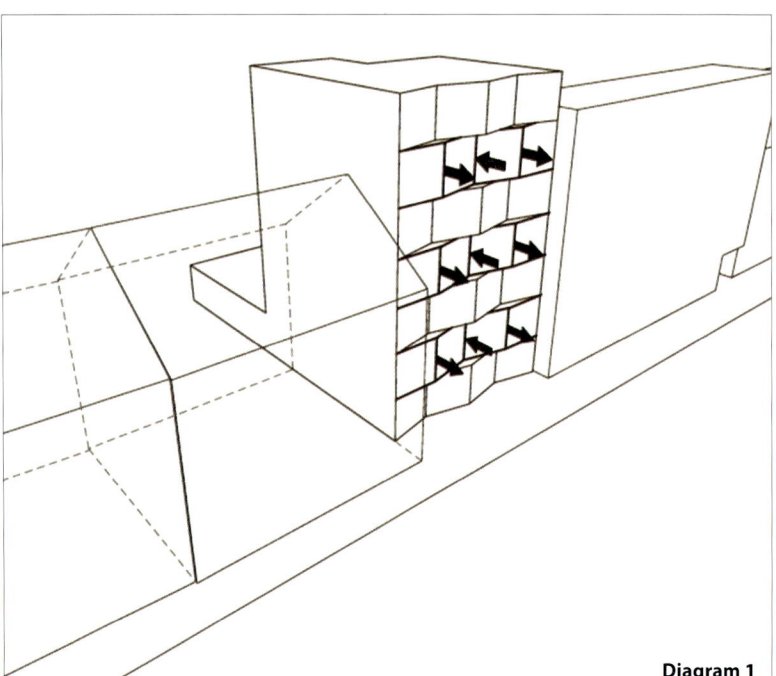
Diagram 1

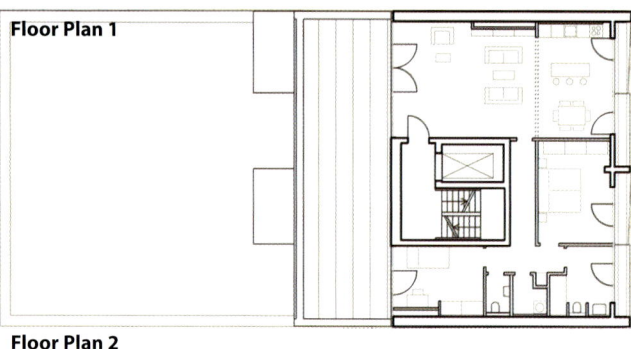
Floor Plan 1

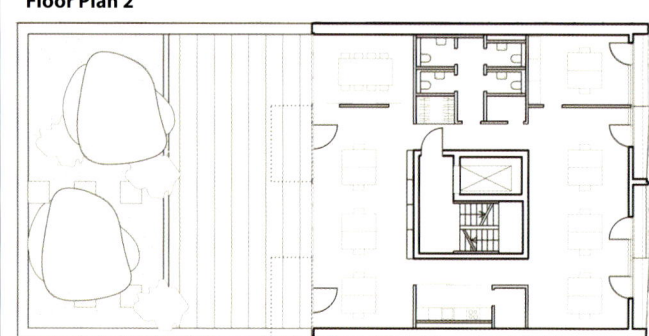
Floor Plan 2

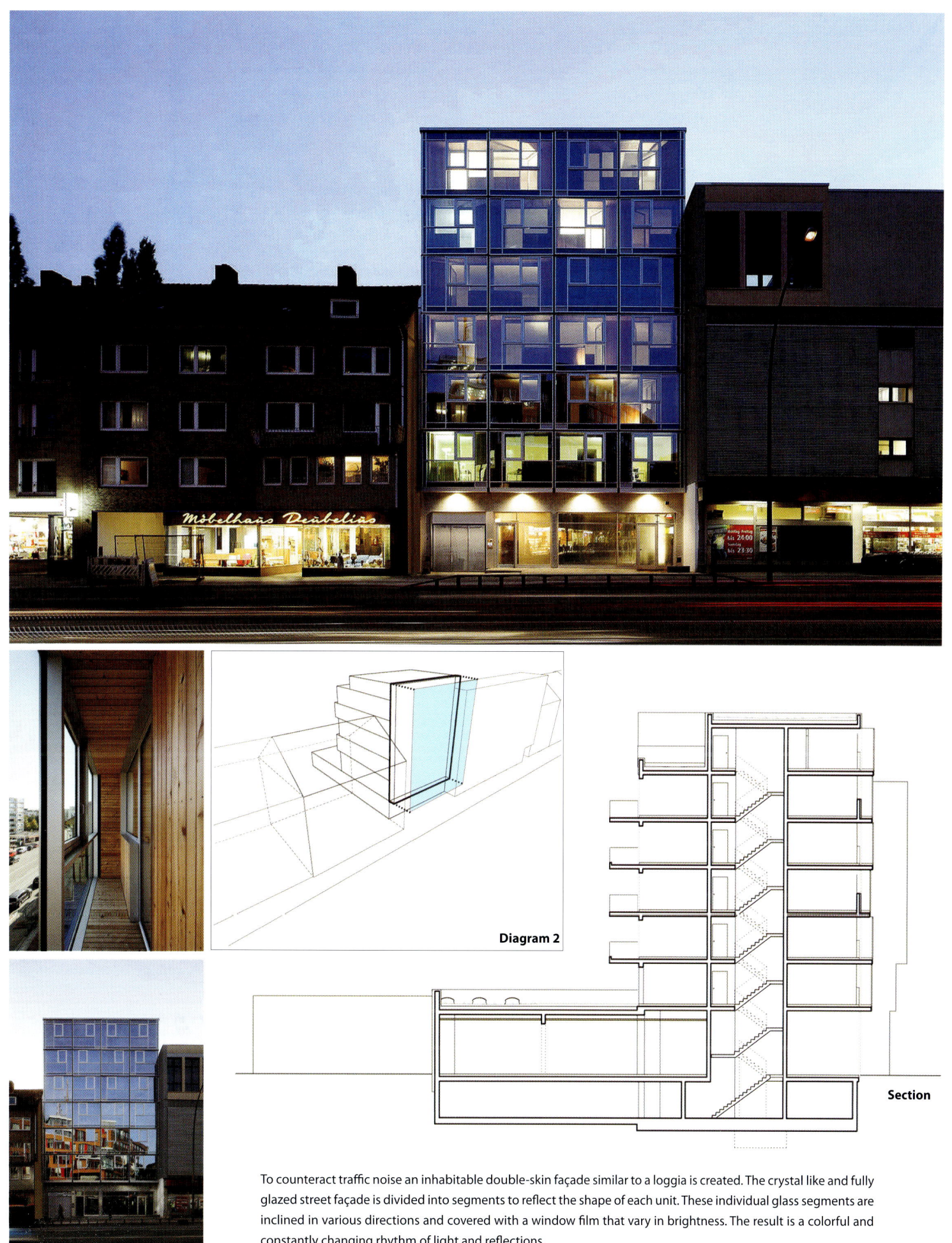

Diagram 2

Section

To counteract traffic noise an inhabitable double-skin façade similar to a loggia is created. The crystal like and fully glazed street façade is divided into segments to reflect the shape of each unit. These individual glass segments are inclined in various directions and covered with a window film that vary in brightness. The result is a colorful and constantly changing rhythm of light and reflections.

Two Tabernacle Street

Architects: Piercy&Company
Location: 2 Tabernacle Street, London EC2A, UK
Area: 1446.0 sqm
Project Year: 2015
Photographs: Jack Hobhouse

Engineer: GDM
Fire Engineer: Exova
Sustainability Consultant: Darren Evans
Planning Consultant: DPP
Project Manager, Cost Consultant: Jackson Coles

Following a fire in March 2010, the burnt out shell of the original building was all that stood on the site of Two Tabernacle Street. The building's owner, Durley Investment Corporation, sought to redevelop the site into offices suited to Shoreditch's media and technology sector. Piercy&Company's response was to approach the challenging L-shaped site as two distinct elements: the re-instatement of the narrow Victorian façade to Tabernacle Street; and brass- clad office spaces in the centre of the site. Details of the surviving features of the Victorian façade to Tabernacle Street were carefully measured and catalogued.

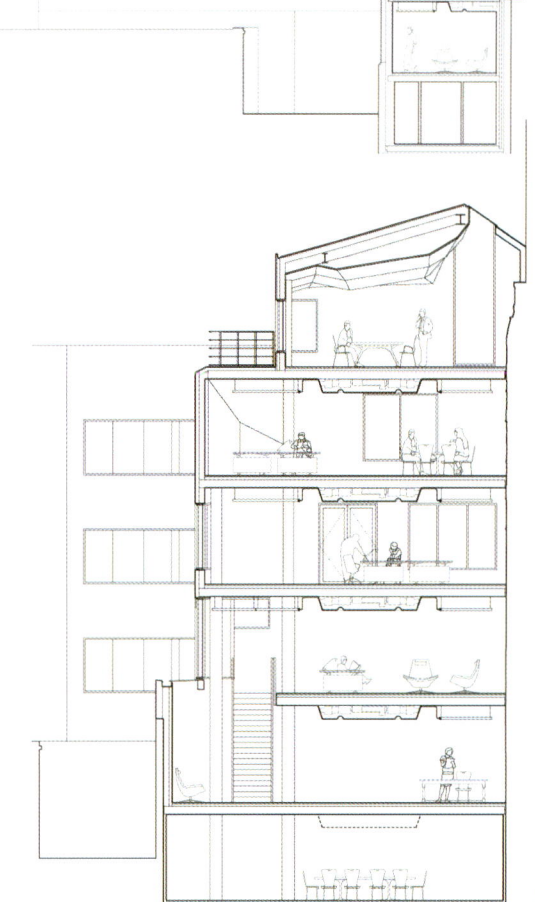

Section 1

Section 2

A refined interpretation of the original period elevation was then painstakingly constructed, to be in keeping with the surrounding Victorian warehouse character and that of the conservation area in which it sits and to align with the revised internal floor levels. Imperial brickwork with weathered pointing and stonework cills, cornices and column capitals bear on one another to create an authentic Victorian masonry frame.

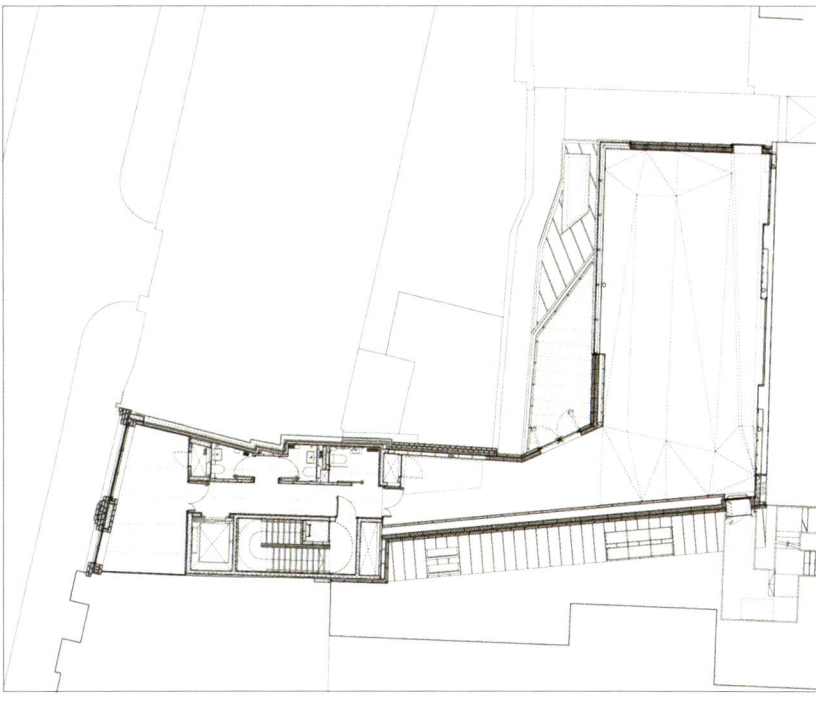

On entering the building, period elements such as plaster cornices, picture rails, a ceiling rose and exposed imperial brickwork in the reception act as a foil to minimalist features, including an angular chandelier (by New York designer Bec Brittain), simply detailed oak wall panelling and large format floor tiles. A generous 6m floor to ceiling height and a small first floor gallery provide an unexpected volume given the narrow street elevation.

Much of the exterior of the building is hidden from the street, only visible to occupiers of the neighbouring buildings. Due to the proximity of existing buildings, rights to light angles carved through the potential building volume. Following many iterations, the final response of cleverly folded planes clad in brass Tecu and punctured by panels of glazing emerged. The use of brass cladding suited the angular, faceted form whilst making subtle reference to the building's industrial heritage.

Maximising natural light in and views out of all the six office floors was challenging on the tight site, further complicated by boundary and party wall matters. Generous lengths of curtain walling were set low to the floor, while strategically placed rooflights flood the floorplates with surprising levels of daylight. At ground level natural light levels are elevated by an atrium and rooflight, illuminating the otherwise windowless space. Throughout the office spaces floor to ceiling heights have been maximised to over 3m by locating services within a central bulkhead and leaving the concrete soffit exposed either side. Stacked above the double height reception, the upper three floors at the front of the building accommodate meeting rooms, which look out over Tabernacle Street through large traditional sash and casement windows. Oak cladding, exposed brickwork and concrete soffits and dark floor tiles make up the material palette of these areas, a gentle contrast to the white walls and expansive glazing of the office floors.

The Newtown School

Architects: Abin Design Studio
Location: Kolkata, West Bengal, India
Area: 15000.0 sqm
Project Year: 2015
Photographs: Ravi Kanade

Design Team: Abin Chaudhuri, Paromita Chatterjee, Poorvi Dugar Ajmera
Structural Consultant: SPA Consultants
Interior Furniture Coordinator: Beautiful Living
Facade Fabricators and Consultants: Annex Design Pvt. Ltd
Signage Fabricator: Ins & Out

This school façade project came to us at a stage when the construction of the 2 academic blocks was already underway. The blocks were rather generic, with 6 floors of classrooms, labs, and other facilities arranged towards the periphery around matching central courtyards. Our brief was to work within these existing parameters to make a school. So we developed a design program in terms of circulation, movement, ventilation, classrooms and other concerns." Keeping in mind the location of the school in the Newtown Area of Kolkata, and its simple surroundings, this school needed to make an impact and establish a distinct identity. The locality is planned in a radial grid and the site for the school is curved along the longer edges. One approaches the site along the inner curve and the blocks are placed at a slight angle facing each other very slightly. They are separated by an active play area. The school also has a swimming pool on its grounds. Of the 2-acre plot, the school occupies approximately 1,60,000 sq ft. of floor space.

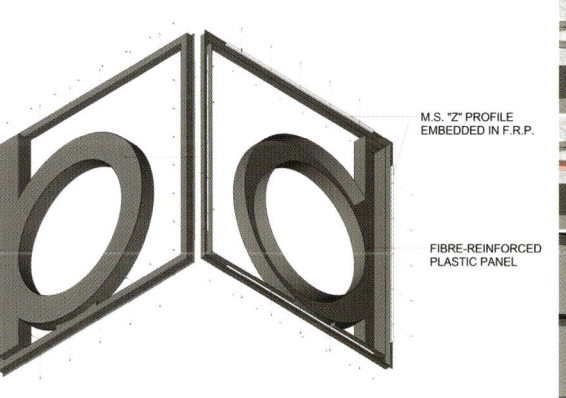

The first step was to create an identity for the school. Our approach was to create a screen that wraps around the buildings and unifies them visually. Also, we intended for the central courtyards of the buildings to merge with the play area creating a seamless connection between junior and senior school. Due to a restriction in program, this ground level connection was not possible. However, the screen was created with a strong character so that its continuous application across the buildings by itself would prove to be a unifying element.

Graphical representations of symbols, alphabets and numbers became an inspiration for the screen. Younger children relate to simple lines as letters of the alphabet and as they grow, abstractions of the same would start to look more like alpha, gamma and pi. Thus, Familiar shapes and symbols were used to create a bespoke stencil screen around the existing unremarkable building mass. The facade not only provides shade to the classrooms from the harsh sun but also lends the school a distinct identity.

488 panels, made of Fibre-Reinforced Plastic (FRP), measuring 3.2 x 3.2 meters envelop each of the two academic blocks. 13 different panels were designed with a combination of small and large alphabets, numbers and symbols. These have been placed in various orientations to achieve a randomized effect on the façade. Structural slabs were projected beyond the building surface all around in a way such that an exact number of panels would fit on all surfaces on the symmetrical cuboids. This also enabled simpler servicing of the panels from the back and ensured a better light quality for the building.

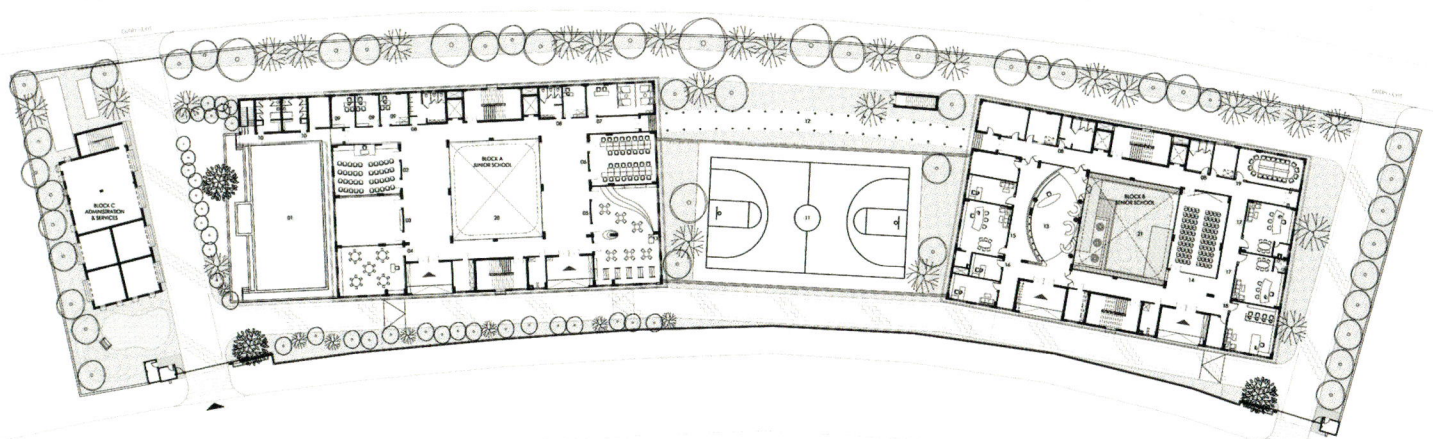

Srithai Super Outlet

Architects: FOS
Location: Chon Buri, Thailand
Area: 4000.0 sqm
Project Year: 2015
Photographs: Teerawat Winyarat

Design Team: Makakrai Jay Suthadarat, Prapaphan Phongklee, Peera Teerittaveesin
Interior Design: FOS [Foundry of Space]
Structural Engineer: Somchai Sakaochanrat
Mechanical Engineer: B.Grimm Trading Corporation Limited
Client: Srithai Superware Public Company Limited

Looking from the main road, the existing 20-year-old depot was once a subdue structure, making no dialogue with the surrounding context and the approaching highway. Due to its strategic location in terms of logistics and urban development in the eastern region, it was unmistakably chosen to be the first own outlet store of Srithai Superware. As a consequence of the selection, the challenges promptly arise; how to turn an ordinary pitch-roof warehouse into an extraordinary attraction amongst target customers and wider public in the region. It also needs to accommodate programmes including an open-plan retail space, display spaces, an adjacent warehouse, a customer service counter, kids club, rental spaces for shops, service areas and an exhibition space of "Srithai Story".

The process of transformation begins with the insertion of new programmes into the existing structural grid system. The existing main structure was mostly kept and strengthened with an exception of the roof tiles being replaced by metal sheet roof with skylight. At the same time, the old facade was stripped off and made way for a new one.

In order to communicate the fundamental character of Srithai products to the public by means of architectural manifestation, the manufacturing process of Srithai plastic products was researched, analysed and then informed the morphological process of the facade design.

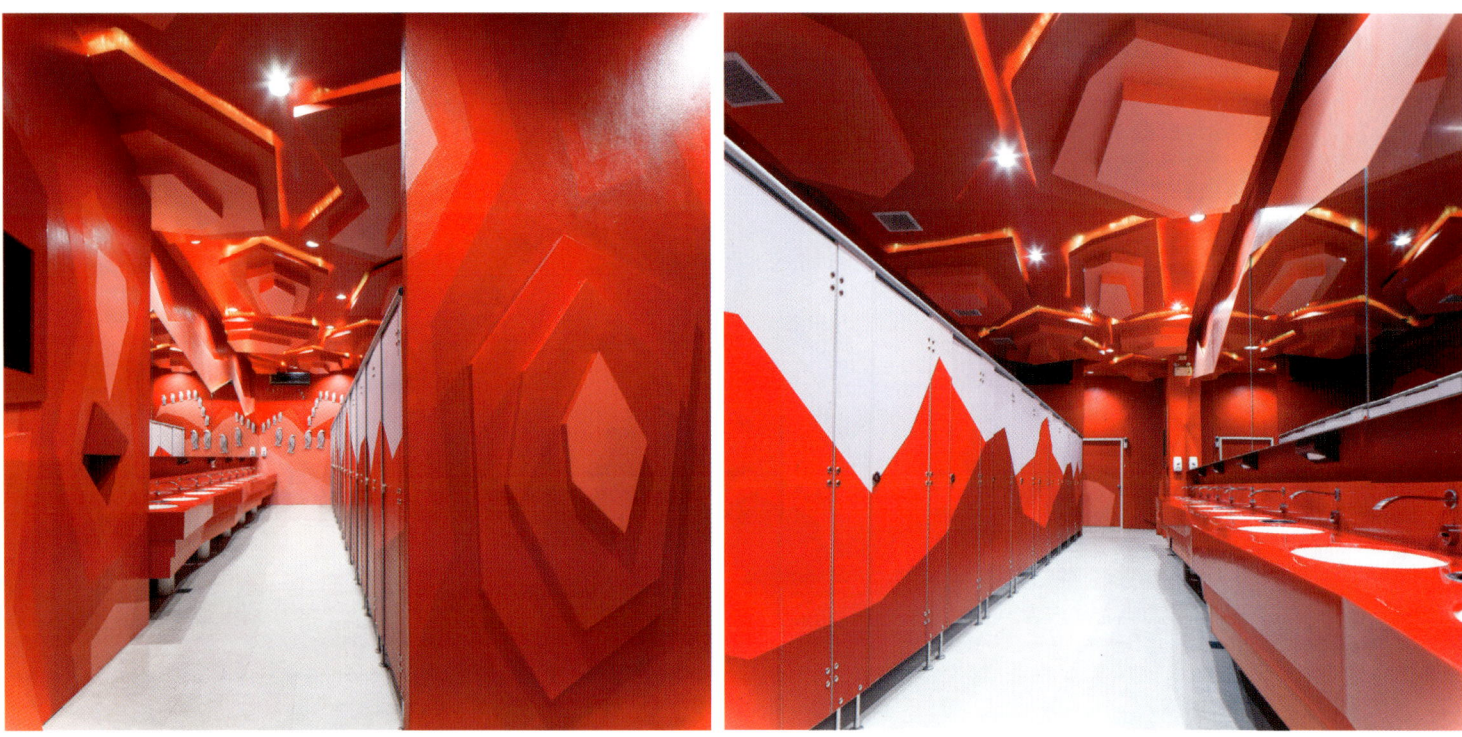

Every curve, shape and form of plastic products is molded from plastic granules, in which they are the smallest elements to form larger, multi-particle entities. Inspired by this transitional stage of plasticity, the extensive front elevation of the building is dissected into series of vertical stripes. These series of stripes then act as a horizontal grid system, in which generic components of folded aluminum panels are inserted. By shifting the next components little by little vertically, the sinuous curves emerge out of the flat plane of the vertical stripes. At the global scale, all the protruded curves flowing along the length of the facade together form familiar silhouettes found along the high stacks of plastic products in the store.

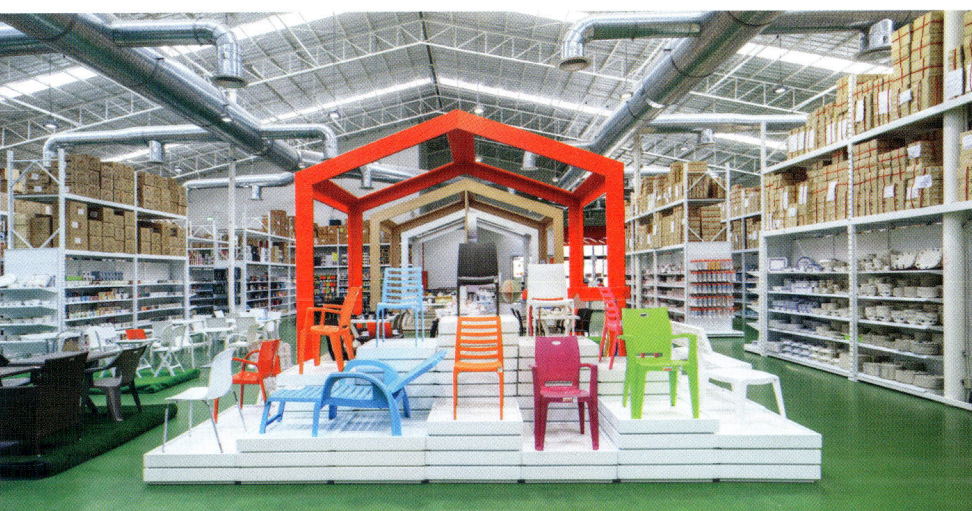

The flow of horizontally stacked components is continued inside, covering the entrance hall of 'Srithai Story', the exhibition space dedicated to the 50-year history of Srithai. At well-located points, certain folded components transform themselves to become display boxes containing trophies, shields and awards achieved by the company. They continually propagate onto the ceiling and become light fixtures, illuminating the foyer area during nighttime.

Scape House

Architects: FORM | Kouichi Kimura Architects
Location: Shiga, Japan
Area: 137.0 sqm
Project Year: 2015
Photographs: Yoshihiro Asada

The house is located in the tiered-developed residential area on a hill. From the site, the beautiful scenery of the lake can be viewed. The customer requested that the view be fully utilized and that the space be opened while not being bothered by eyes of neighborhood. In this project, versatile spaces that incorporate light and scenery were intended by the windows in order to bring out the best in this house. Scenery viewed through a window is greatly affected by the size or position of the window.

Section

Ground Floor Plan

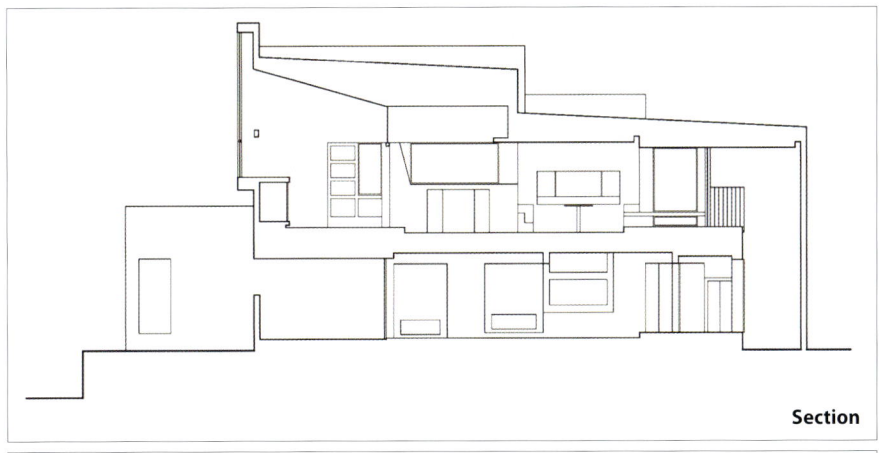

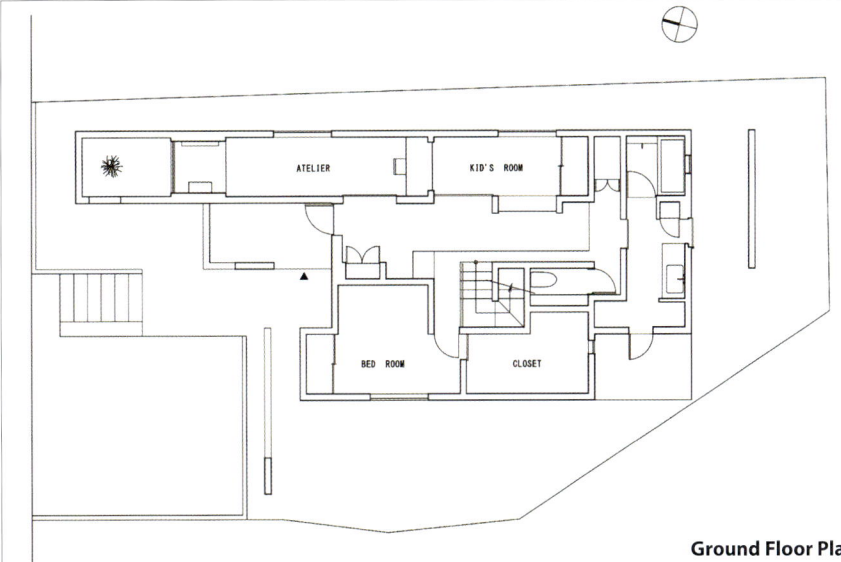

232

First Floor Plan

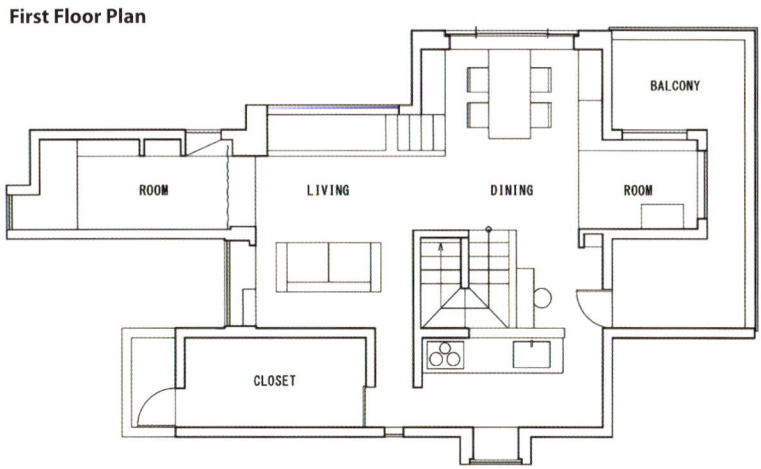

It is therefore essential to consider what should be viewed or not in the scenery framed by the window, instead of being stereotyped to take in the large area of the scenery by providing the largely-opening window. The windows as framings produce comfortable spaces where you can enjoy light and scenery without being annoyed by eyes of neighborhood. The spaces incorporate a table, bench, book shelf, niche, and other furniture items so that you can utilize there to view outside, read books, eat meals, etc., which brings out characteristics of each space and provides its versatility. The space is composed of mortar with a feel of texture, highlighting its presence. At the same time, it provides openness created by the clear and continuous sightline.

The dynamic configuration involving the box-shape volume with rhythmical layout of the windows produces beautiful life scenes where light and scenery are taken in while the eyes of neighborhood are blocked.

Sonnenhof

Architects: J. MAYER H. Architects
Location: Sonnenhof 9, 07743 Jena, Germany
Area: 9555.0 sqm
Project Year: 2015
Photographs: David Franck

Design Team: Christoph Emenlauer, Jesko Malkolm Johnsson-Zahn, Christian Paelmke
Project Architects: Jens Seiffert
Project Management: Kappes Partner IPG GmbH, Berlin
Inspection Engineer: Dr.-Ing. Joerg Diener, Erfurt
Interior Construction Office: Friedrich Wackerhagen GmbH&Co.KG

Sonnenhof consists of four new buildings with office and residential spaces. Located on a consolidated number of smaller lots in the historical center of Jena, Germany, the separate structures allow for free access through the grounds. Their placement on the outer edges of the plot defines a small-scale outdoor space congruent with the medieval city structure. Its outdoor facilities continue the building's overall design concept past the edges of the lot. The planned incorporation of commerce, residence, and office enables a flexible pattern of use that also integrates itself conceptually into the surroundings.

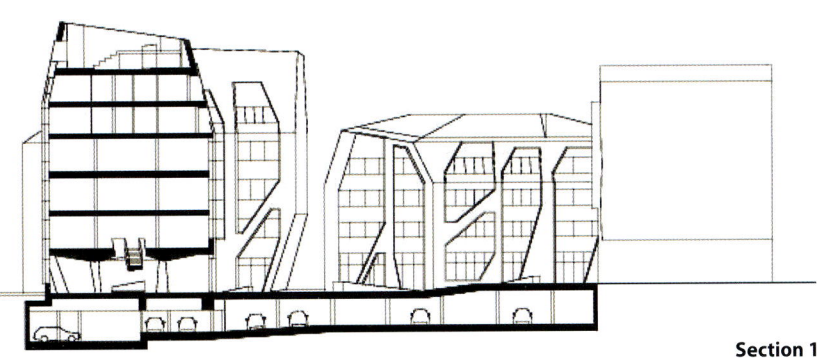
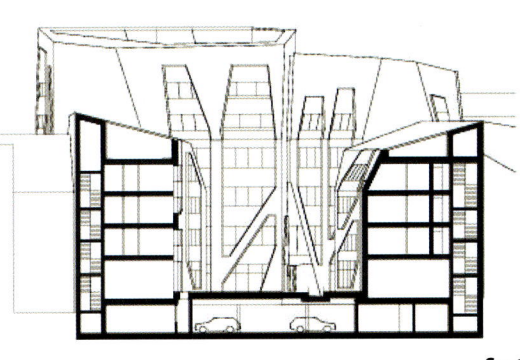

Section 1

Section 2

236

Renovation of México Fortius Office Building

Architects: ERREqERRE Arquitectura y Urbanismo
Location: Avenida Río Consulado 114, 7 de Noviembre, 07840 Mexico City, Mexico
Project Architect: Rafael Ponce
Project Area: 1697.0 m2   **Project Year:** 2015
Photographs: Onnis Luque

The project is the renovation of an existing building, which was resolved respecting the main structure and its 5 levels, adapting the building for commercial use on the ground floor and offices on the upper 4 levels.

The main design guidelines were maximizing the existing conditions of the property, including the use of a courtyard that allows light and ventilation inside the building, which is presented as the heart and hinge of the space inside.

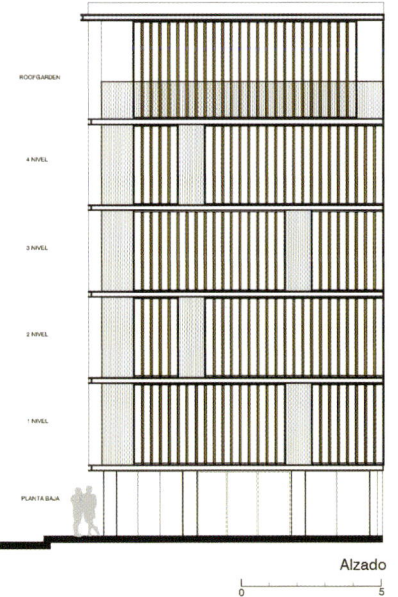
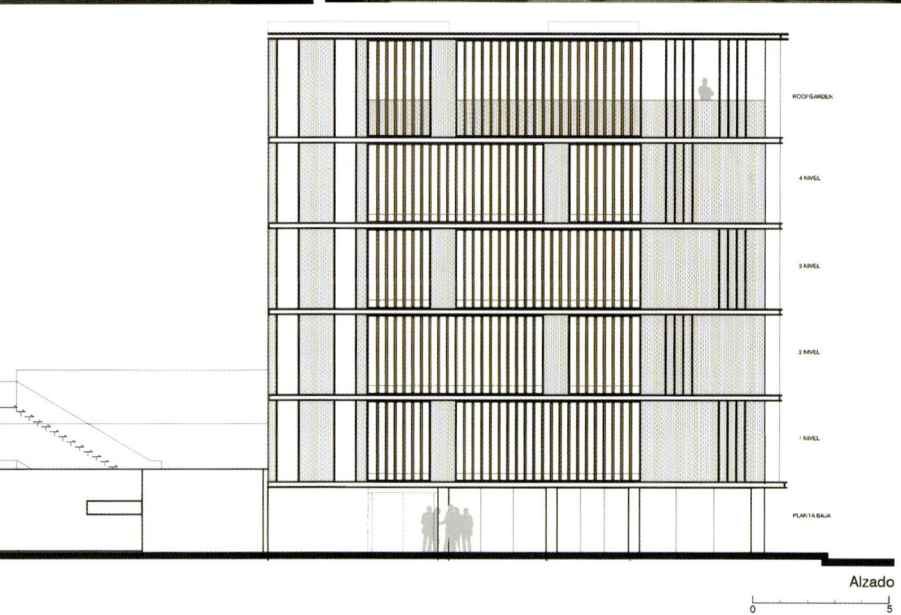

Its orientation and location was key to solving the exterior design of the facade, with a South-West corner and facing a main road as Rio Consulado, we use a system of operable vertical sunbreakers to allow the passage of sunlight in a controlled manner, also helping to mitigate noise and control the views of the road.

These solutions allow us to give the project sustainable conditions, designing comfortable spaces for their respective uses inside. In the architectural applications, the solution to the mechanical control of the facade has been provided by our partner Somfy.

Naz City Hotel Taksim

Architects: Metex Design Group
Location: Dolapdere Cd, İstanbul, Turkey
Area: 500.0 sqm **Project Year:** 2015
Photographs: Cemal Emden
Design Team: Kağan Erk, Sinan Kafadar
Interior Design Project: Kağan Erk, Burcu Arkut
Contractor: Regnum Turkey
Mechanical Project: Vemeks Engineering
Electrical Project: Key Engineering
Interior Finishings: Layık Decoration

The hotel is located at the main district of Dolapdere going to Taksim. It is the new developing area. There are other hotels as well as the Bilgi University and the Koç Museum that is under construction (to be completed in 2016). It is the meeting point of the tourists, artlovers, businessmen.

The architecture of the hotels is related with the hotel's environment. As it is in the middle of a developing area of Istanbul, we wanted to show its difference from the other buildings around it. However, we did not want it to be in contrast with them. Therefore, the enviorment's variety, irregularity and randomness is interpreted to provide it in the design of the hotel.

It is seen in the facade with the same square zinc metal element to be used at each rooms facade. The square's distorted faces change as the main square shape placed for each of the room. The need of either a divan or armchairs at the interior of the particular room is decided accordingly. The movement on the facade is increased by the reflections of the sun on the zinc surfaces throughout the day.

The artworks that we have been chosen for this hotel, are also inspired by the idea of these three terms: variety, irregularity and randomness. In the rooms, we have decided to place a 3D artwork. The shapes were irregular and matched with the general colors of the executive rooms. The artworks in the executive rooms were not supposed to stand out. They should give the feeling of being a part of the wall panels. For the suit rooms, the choice of artwork technique were rather different. We would like to see more color in these room types for taking the rooms to another level.

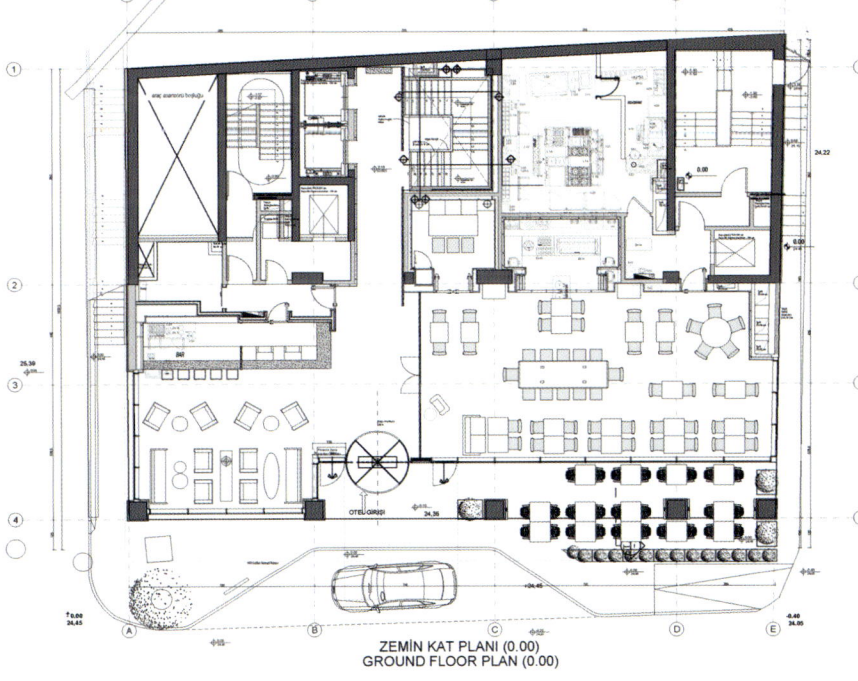

ZEMİN KAT PLANI (0.00)
GROUND FLOOR PLAN (0.00)

Nature & Environment Learning Centre

Architects: Bureau SLA
Location: Heggerankweg 871, 1032 JC Amsterdam, Netherlands
Design Team: Peter van Assche, Ninja Zurheide, Ane Arce Urtiag, Joti Weijers Coghlan
Area: 281.0 sqm   **Project Year:** 2015
Photographs: Filip Dujardin   **Cost:** € 400.000 ex. VAT

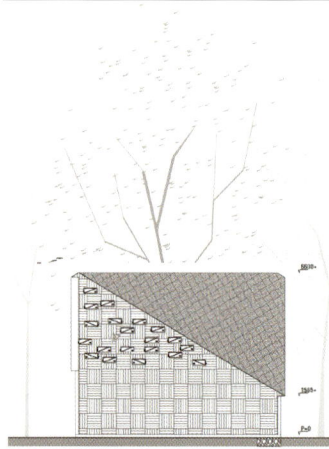

The location of the NME, situated in the middle of the school gardens of Amsterdam Noord, was just about perfect: a rectangular plot near the entrance of the site, optimally oriented to the gardens. The only significant challenge was that the long facade of the proposed structure was not precisely south-facing, which meant that the building's solar collectors would not have been optimally placed for collecting the sun's rays. But by positioning the ridge of the roof exactly along an east-west axis - offset from the walls of the building - two things were accomplished: firstly it solved the solar collection issue by ensuring the solar collectors were precisely south-facing, and secondly it provided the building with its very distinctive shape. The design of the Centre's roof also provided another learning opportunity for students due to its close proximity to the ground: at the roof's lowest point even small children can easily see the solar panels. Inside the building, the floor plan is almost symmetrical, with the entrance in the middle. On the left and right are classrooms of equal size and on the first floor two identical rooms serve as a office space and a canteen. In the middle there is nothing, which provides a view through the glass entrance doors straight through to the school gardens on the far side of the building.

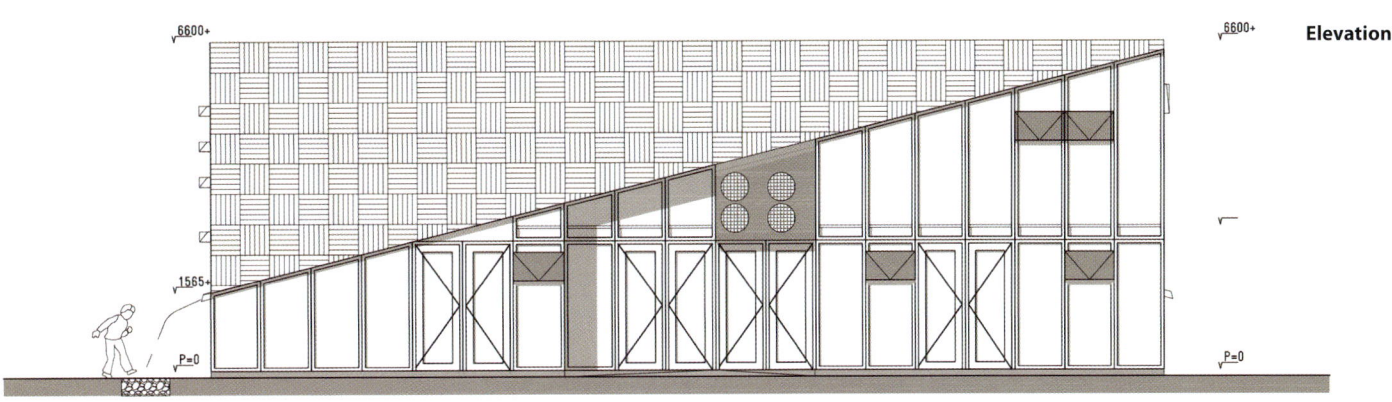

Elevation

Naman Spa-Vietnam

Architects: MIA Design Studio
Location: Da Nang, Da Nang, Vietnam
Architect in Charge: Nguyen Hoang Manh
Area: 1600.0 sqm   **Project Year:** 2015
Photographs: Oki Hiroyuki

Conceptual Design: Nguyen Hoang Manh, Nguyen Quoc Long
Technical Design: Bui Hoang Bao
Interior Design: Steven Baeteman, Truong Trong Dat, Le Ho Ngoc Thao
Developer: Thanh Do Investment and Construction Cooperation

The Pure Spa is an oasis of tranquility and facilitates the five-star Naman Retreat, Danang. Fifteen stunning treatment rooms are endowed with lush open air gardens, deep soak bathtub and cushioned daybed built for two. Keep fit at the equally sleek health club with gym, meditation and yoga sessions held at the open lounge garden in the still cool mornings. The ground floor contains open spaces with relaxing platforms surrounded by serene lotus ponds and hanging gardens. A true space where all senses are touched and the mind comes to peace…

The architectural design company MIA Design Studio's ingenious use of natural ventilation keeps the building cool and gives the guest a refreshing experience. With use of local plants, each retreat becomes a healing environment where the guest can enjoy a luxurious wellness in privacy.

ELEVATION

SECTION 01
0   5m

250

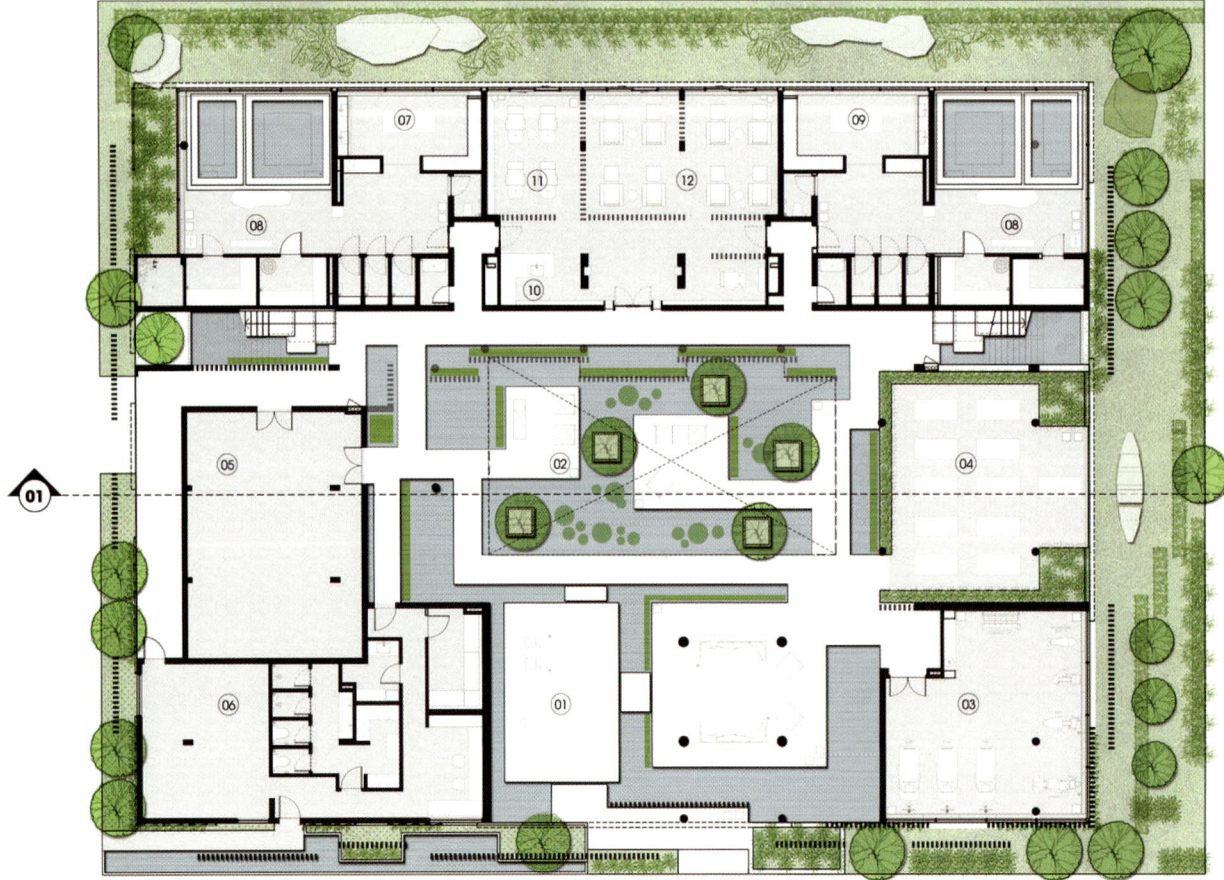

GROUND FLOOR

01. Lobby area
02. Cabana
03. Gym room
04. yoga area
05. Function room
06. Staff room
07. Locker female
08. Jacuzzi
09. Locker male
10. Pantry
11. Shampoo area
12. Relax area

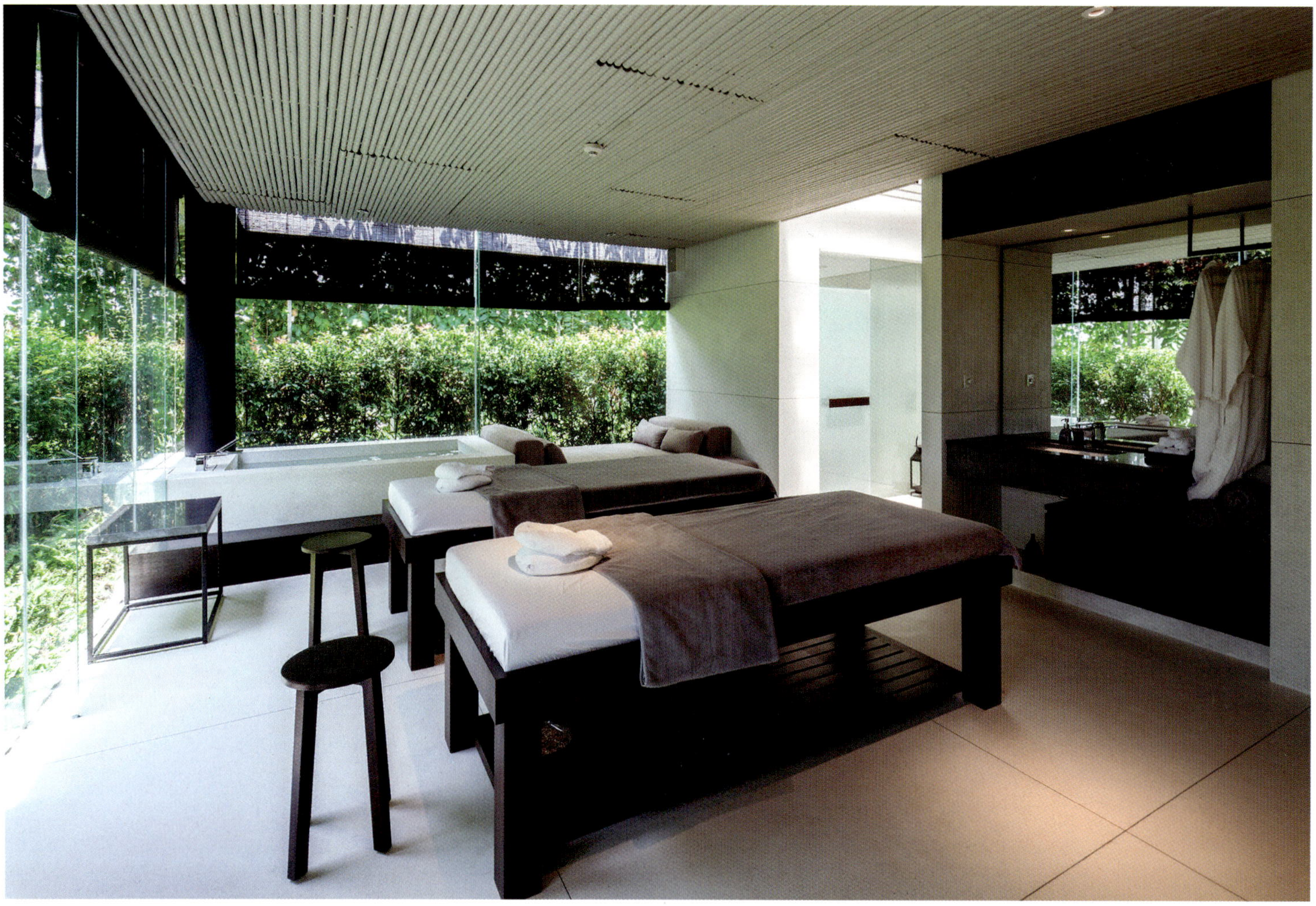

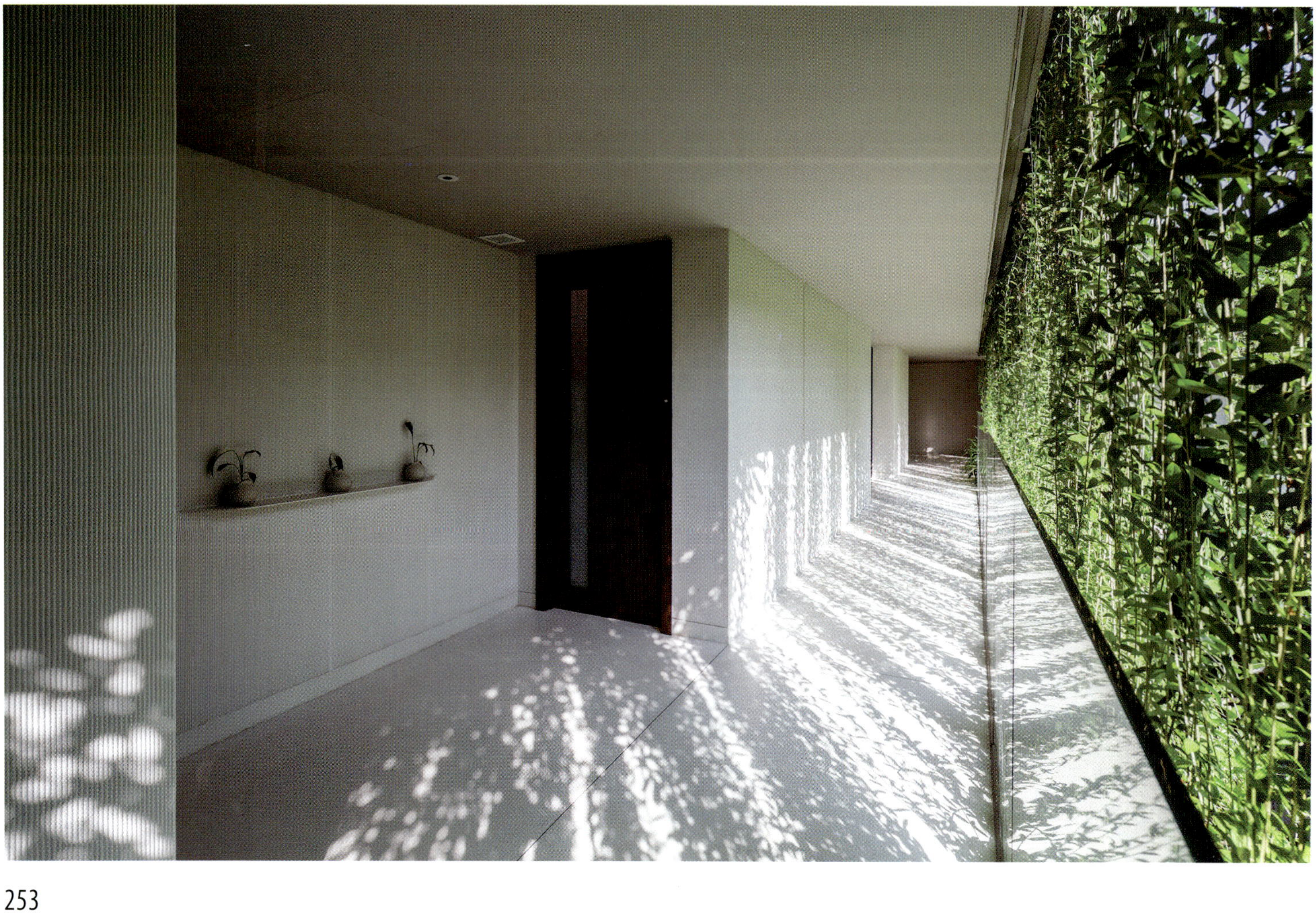

Different area's flow smoothly into each other and the beautiful landscape creates an amazing journey into a dream like experience. The facade is composed by lattice patterns alternated with vertical landscapes that filter the strong tropical sunlight into a pleasant play of light and shadow on the textured walls. Various plants are carefully allocated and become a part of the architectural screens.

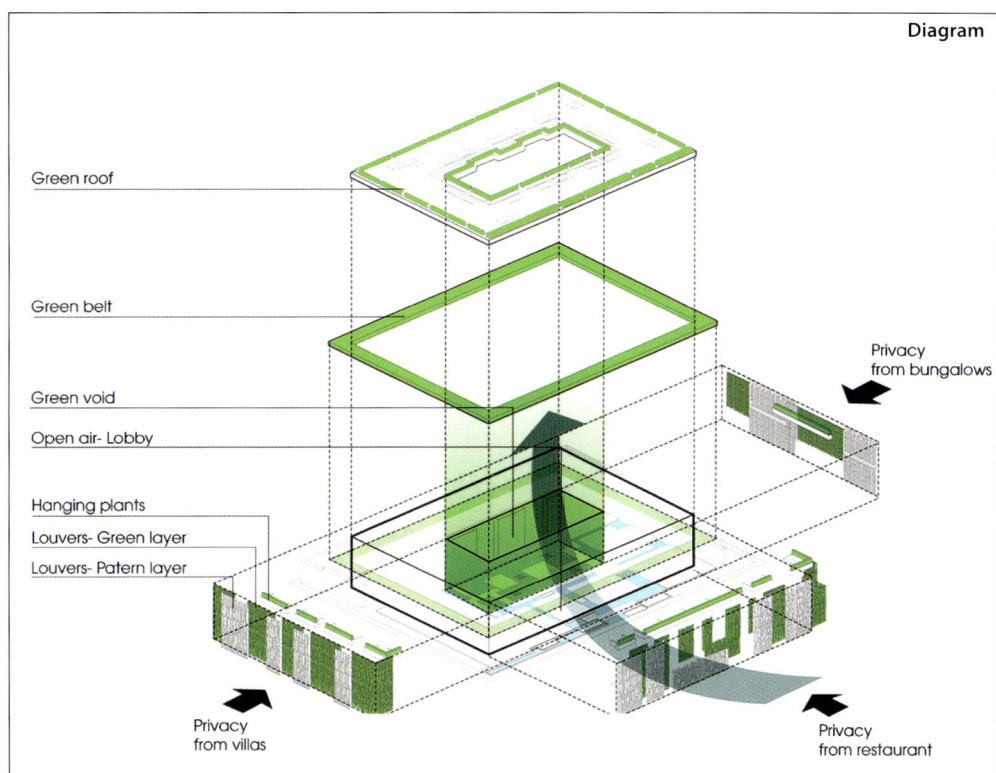

Diagram

Museum for Qujing Culture Center

Architects: Hordor Design Group, Atelier Alter
Location: Qujing, Yunnan, China
Area: 202361.0 ft2   **Project Year:** 2015
Photographs: Courtesy of Atelier Alter
Project Architect: Qiuda Lin, Weining Lin
Architect in Charge: Yan Huang
Design Architect: Yingfan Zhang, Xiaojun Bu
Architecture Design Team: Zhanghan Zheng, Zhenqing Que, Ling Zeng
Structure: Masha Yang, Yingying Lin
MEP Engineer: Guangyu Zhang, Xiaonan Zhu, Xiaoan Han

Two impossible miracles co-exist in city of Qujing: the Longyan Tablet and a fish fossil of 4,000-billion-year-old. While Longyan Tablet marks the invention of a prominent calligraphic style, the fish fossil rewrites geology in human history. The archaeological relics are both metaphor and subject matter of the project. According to Plato, "whatever once exists can never cease to exist," the collective consciousness of the city and her citizens awaits a resurrection, in a contemporary setting, through the materialization of a series of significant projects.

The museum is entered through center of building mass. Audiences are elevated to a concrete plateau, as they begin their exhibition routes at a strategic point in space. A vertical plaza is defined by processional steps and its echoing suspended roof. The graduate suspension of the enormous roof presents an "anti-gravity" architecture statement that puts the audiences in awe.

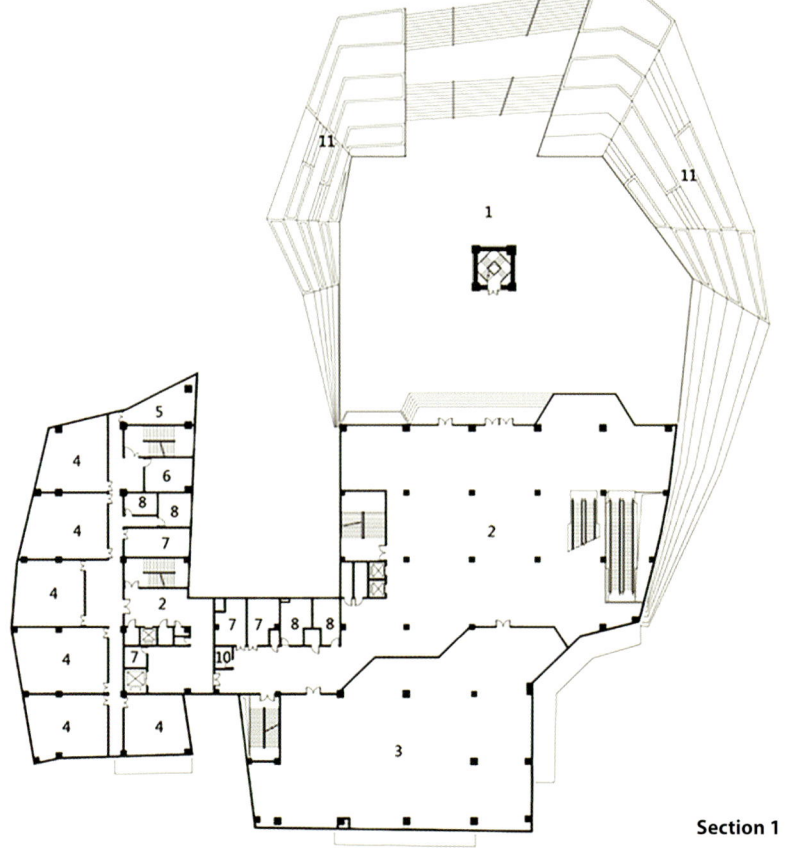

Section 1

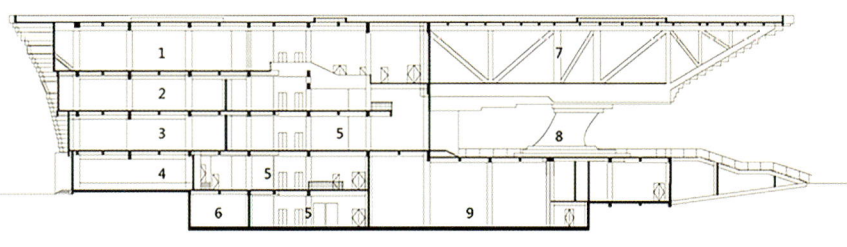

1. Cuan Cultural & Non-material cultural exhibition hall
2. Exhibition hall
3. Natural biological specimens、Human prehistory exhibition hall
4. Multifunctional lecture hall
5. Hall
6. Equipment
7. Folk cultural relics exhibition hall
8. Entrance platform
9. Temporary exhibition hall

Section 2

One could experience the immensity of space and time the city occupies. The strong presence of the void reinstates the gravitas of the museum's subject matter: a profound history dates back over 4,000 billion years.

A vertical landscape made of concrete drapes down from roof to the ground. As audiences penetrate the landscape and reach to the exhibition zone, geography and humanity converge at that every moment.

Instead of assimilating into analogies of the site—terrace field, fossil grain, or calligraphic strokes—the formal expression of the architecture is in dialogue between the concrete and the abstract, the familiar and the unfamiliar.

Montforthaus in Feldkirch

Architects: HASCHER JEHLE Architektur
Location: Montfortgasse 1, 6800 Feldkirch, Austria
Project Head: Frank Jödicke, Gorch Müllauer, Markus Mitiska
Project Year: 2015
Photographs: Svenja Bockhop, Benjamin Marte

Interior Design: Mitiska.Wäger Architekten + HASCHER JEHLE
Light planning: LDE, Eschen (Liechtenstein)
Floor Area: 10.840 qm
Total Area: 13.435 qm

The new Montforthaus is a multi-purpose cultural centre for the people of Feldkirch and the surrounding region. It is versatile enough to host conventions, balls, trade fairs, classical concerts, pop concerts and theatre performances. The Montforthaus is harmoniously embedded in the historical urban grain of the medieval old town of Feldkirch. While its formal articulation is demonstratively modern, its materiality picks up the traditional Jura marble of the region, setting up a dialectical frisson between the two while simultaneously weaving the new insertion into the existing fabric of the town.

Like a pebble in the riverbed of the town, the new cultural centre sits in the flow of urban space between three adjoining squares which fuse into a single large urban space. The same natural flow of space continues into the building, leading visitors into a four-storey landscape of foyers and open galleries beneath a naturally illuminated glazed atrium roof. With its transparent front the Montforthaus is inviting passers-by into the Montforthaus. It leads directly into the almost 15 metre high, brightly lit open foyer. The seamless flow of space from outside to inside and the fully glazed walls of the foyer make the surroundings part of the space and contribute to the sense of an expansive interior. A broad sculptural stair leads visitors up from the entrance to the gallery levels, the small auditorium, seminar rooms and from there on to the roof terrace.

Section 1

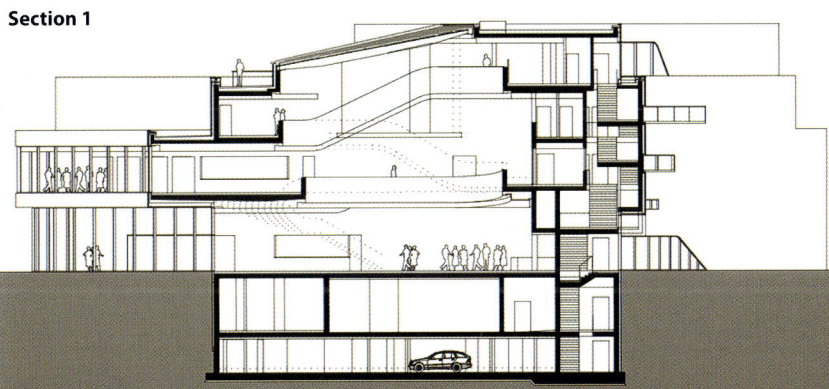

Floor Plan 1

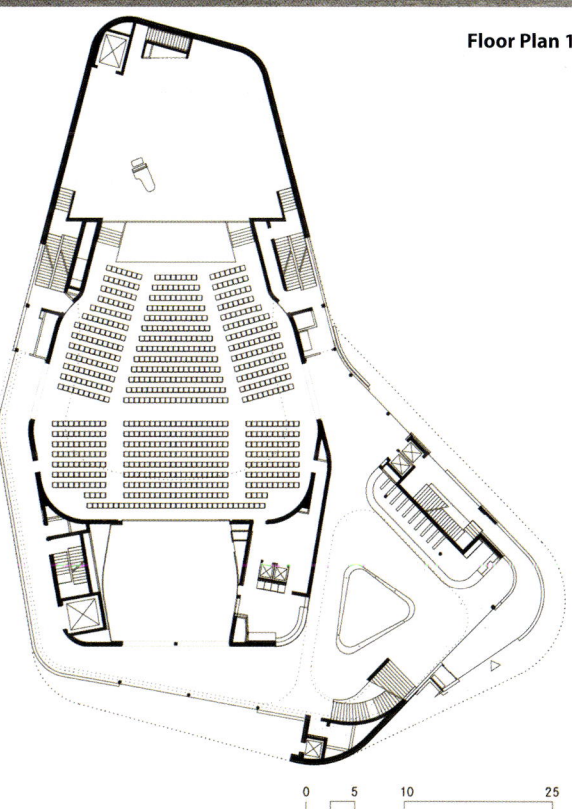

The vertical patterning of the natural stone façade made of light Jura limestone underlines the sculptural form of the building without requiring curved sections and lends the building a sense of elegant restraint. The Jura limestone façade, comprised of 30 cm wide, 4 cm thick and at times over 200 cm long hung stone panels, has a total surface area of 2150m^2.

Section 2

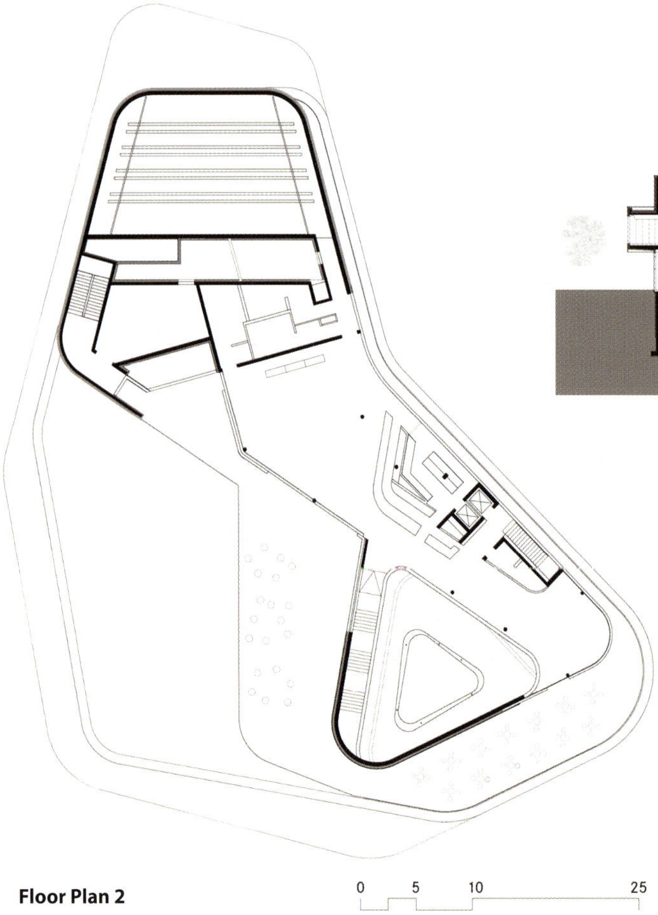

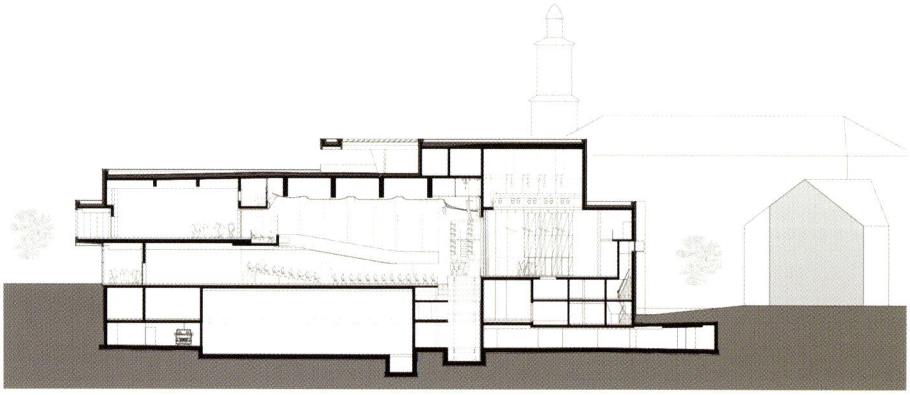

Floor Plan 2   0 5 10   25

The vertical panel edges are precisely milled to allow them to overlap, obviating the need for open panel seams. The Montforthaus has been planned in accordance with "Green and Smart Building" concepts. During all stages of the design process, opportunities to exploit synergy effects between the architecture, technical services and building physics were investigated. The integral life cycle concept reduces both the investment as well as the running costs. Coatings, paints and insulation materials, as well as the façade materials, were selected and used according to environmental criteria. Parking spaces for electric bicycles and charging stations for electric vehicles complement the sustainability credentials of the building.

Miu Miu Aoyama Store

Architects: Herzog & de Meuron
Location: Tokyo, Japan
Area: 269.0 sqm
Project Year: 2015
Photographs: Nacasa & Partners

Design & Build Contractor: Takenaka Corp., Tokyo, Japan
Landscape: Vogt Landschaftsarchitekten, Zürich, Switzerland

We already noticed this over 10 years ago when we were planning the glass building for Prada Aoyama. At that time, we were interested in counteracting the situation – on one hand, by placing a small plaza to the side of the building, and on the other, by making the structure completely see-through so that one can see into the interior from all sides and can also look out from inside at specifically targeted views of the city.

Over the past decade, the distinctive building has become a much-frequented location and it was therefore important to Prada, our client Prada Japan and also to us as architects to take this into account in planning the Miu Miu store located in the immediate vicinity on the opposite side of the street. We started out by trying several different architectural typologies. Since zoning regulations called for less height, we explored the potential of a smaller, more intimate building. We used the following thoughts to channel our ideas: more like a home than a department store, more hidden than open, more understated than extravagant, more opaque than transparent.

The façade has neither logo nor pomp; it is a polished, mirror-smooth surface, as if one single giant brushstroke had swept smooth the ordinarily matte surface of the steel panelled façade. This surface attracts the gaze and curiosity of passing pedestrians. But instead of affording a view inside, as in a shop window, the gaze is inverted; instead of the anticipated see-through window, viewers encounter self-reflection.

While the street is not a place that encourages lingering and looking around, the building itself is a gesture that extends an invitation to come inside and stay a while."

The typological model that best suited these considerations and specifications was a box placed directly at the level of the street, its cover slightly open to mark the entrance and allow pedestrians to look inside. Only then do they realize that the building is a shop. Here, under the oversized canopy, the two-storey interior is visible at a single glance, as if the volume had been sliced open with a big knife, turning the inside out. The rounded, soft edges of the copper surfaces inside meet with the razor-sharp steel corners on the outside of the metal box, while the cave-like niches clad in brocade face the central space of the shop like loges in a theatre. The shop on two tall storeys not only presents enticing goods on tables and in display cases; it is also like a spacious and comfortable home with inviting sofas and armchairs.

Mariehoj Cultural Centre

Architects: Sophus Sobye Arkitekter, WE Architecture
Location: Holte, Denmark
Project Year: 2015
Photographs: Rasmus Hjortshoj COAST Studio, Courtesy of Lars Eldrup

Design Team: Sophus Søbye, Elena Ardighieri, Maria del Mar Freire Morales, Jan Nielsen, Ejvind Christiansen, Marc Jay, Julie Schmidt-Nielsen, Thomas Helsted, Kurt Ohlenschlaeger, Hermanus Neikamp, Lawrence Mahadoo, Lena Reeh Rasmussen, Jenny Sellden, Zsofia Horvath, Nora Fossum

Mariehøj Cultural Centre draws a clear profile in the landscape in Holte, Denmark. The new foyer invites all people in and functions as a heart bringing people together and highlighting the many users and activities in the house.

The cultural centre merged together with the green landscape bridging the gap between the arrival area, the cultural plaza and the beautiful backyard of Mariehøj. The building opens up towards the surroundings and incorporates the green qualities to the activities in the house.

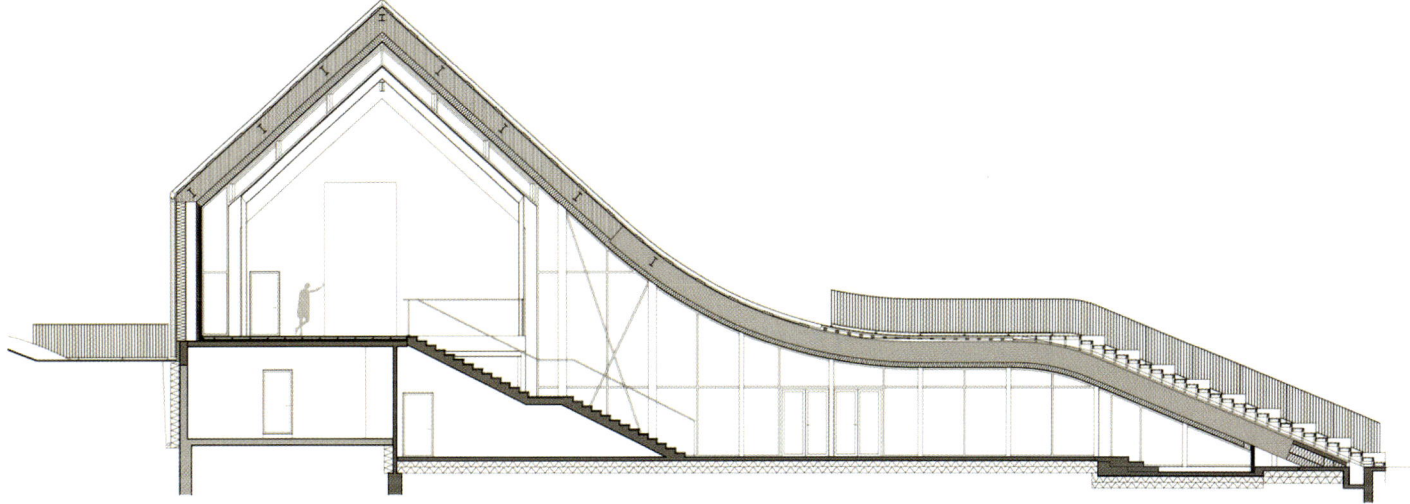

Kulturcenter Mariehøj, Længdesnit, Mål 1:150

The house is both a cultural activity centre and a well-functioning working place. Through reorganization and rebuilding, better spaces for individual activities and appropriate positions of the different functions are created. At the same time more meeting points are created, which bring the multiple activities in the house together and offer a space where new meetings and activities across interest and age arise.

The new cultural centre is a single organism with a high aesthetic and functional quality. The reconstruction of the house is not just an upgrade but also a vision for how the cultural centre can meet future challenges and enhance the unique character and common stories.

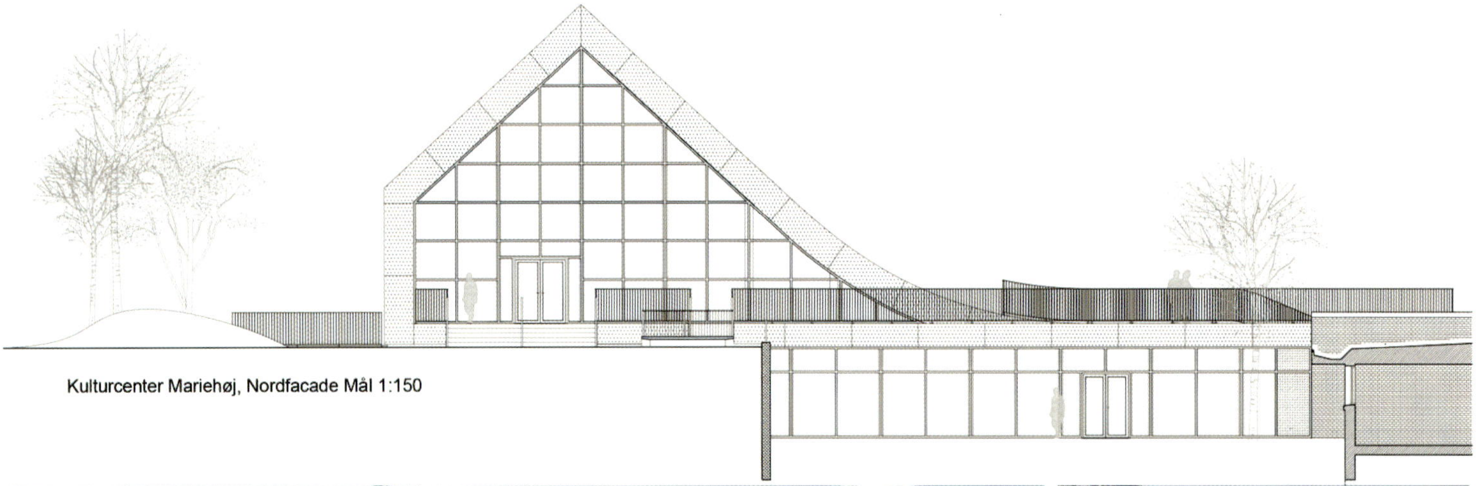

Kulturcenter Mariehøj, Nordfacade Mål 1:150

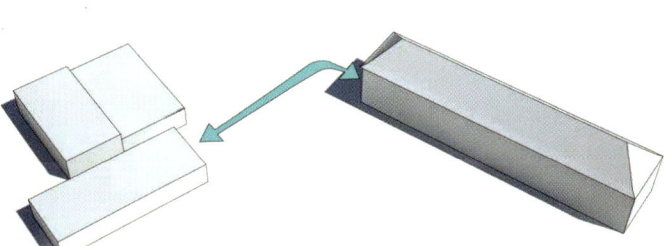

En ny foyerbygning forbinder og samler kulturcentrets bygningskroppe.

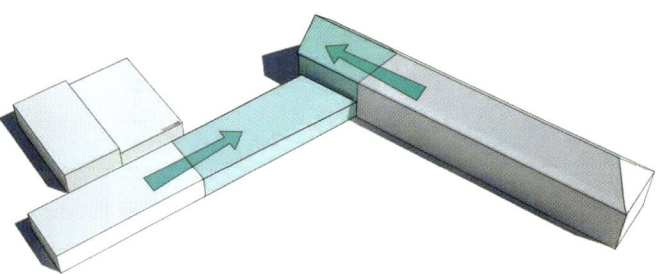

Bygningerne ekstruderes ud så de forbinder både i 1. og 2. sal

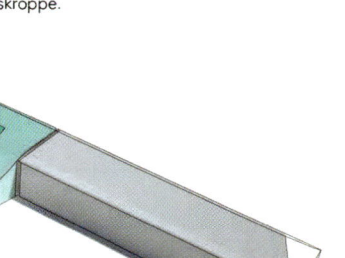

De to nye bygningskroppe forbindes og sammensmelter i én form.

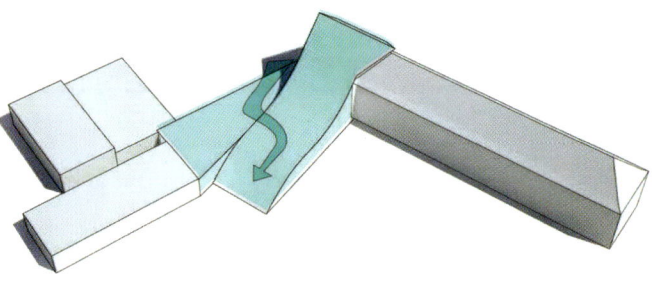

Bygningen splittes og foldes ned, så der skabes en udendørs forbindelse mellem for- og bagside over bygningen.

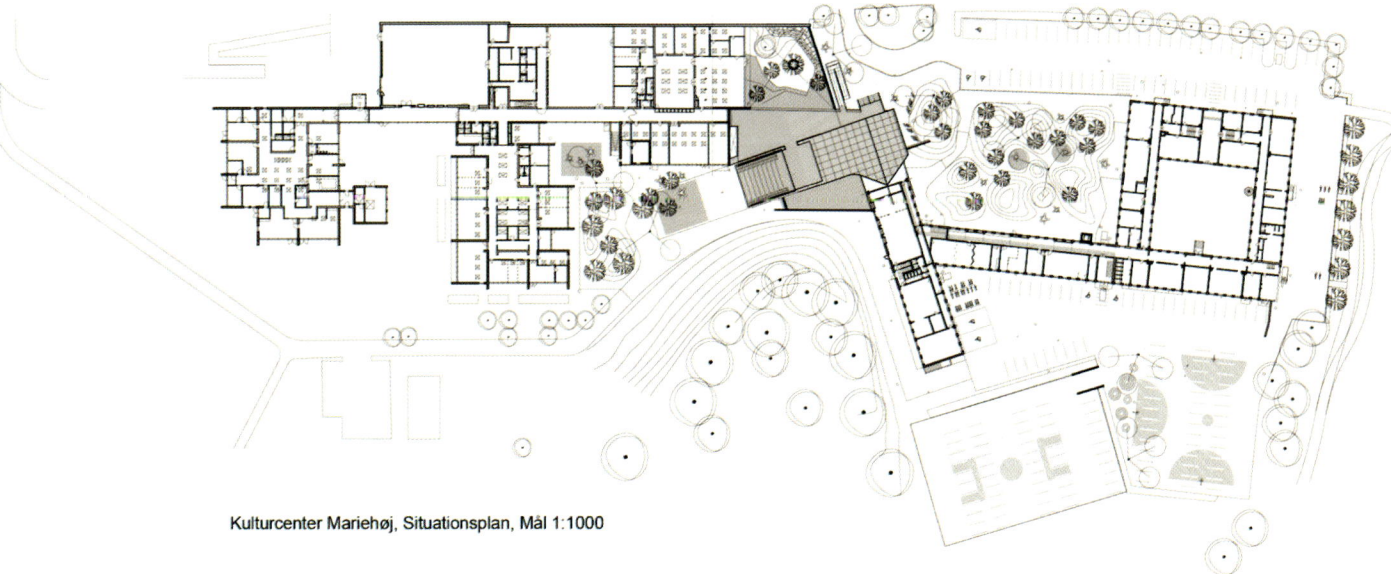

Mariehøj Cultural Centre invites people to draw inspiration from new stories, complete experiences and the cultural life. The cultural centre inspires and generates both familiar and new interests, and promotes new meetings between users and staff. The new foyer takes the best of the existing buildings styles and introduces a new typology not foreign to the rest of the building.

Kulturcenter Mariehøj, Situationsplan, Mål 1:1000

274

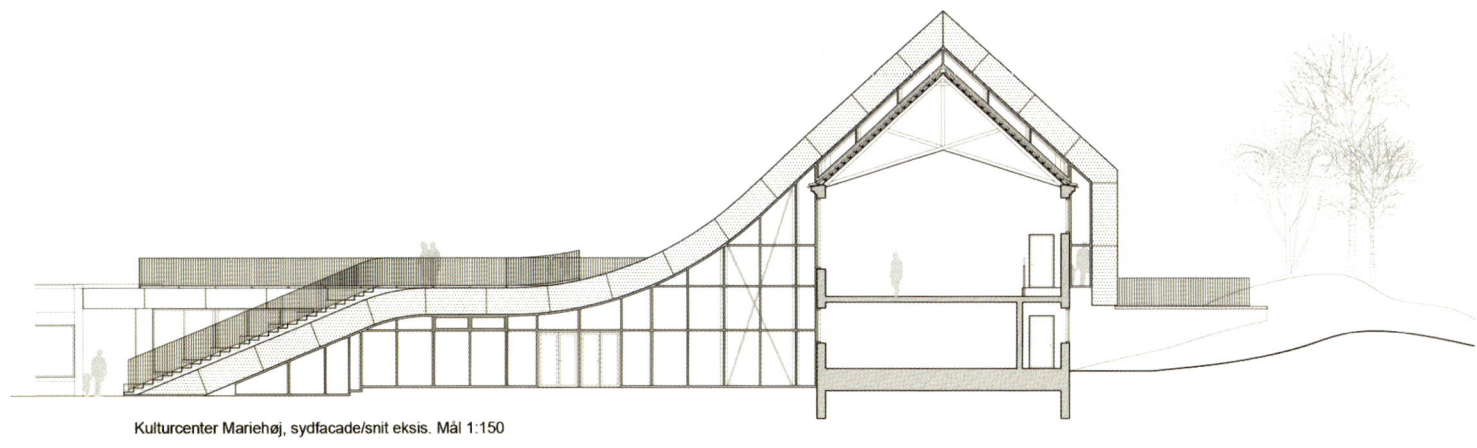

Kulturcenter Mariehøj, sydfacade/snit eksis. Mål 1:150

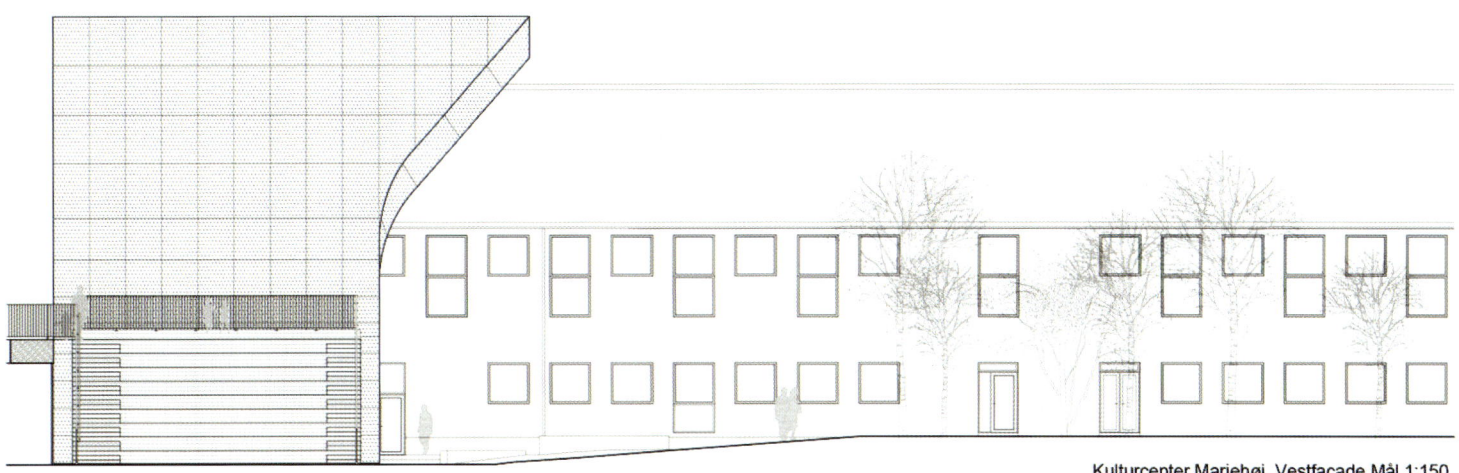
Kulturcenter Mariehøj, Vestfacade Mål 1:150

Library and Culture Centre

Architects: Primus Architects
Location: Copenhagen, Denmark
Collaborators: Rytter A/S, JPM engineering, GHB Landscape architects
Area: 2200.0 sqm   **Project Year:** 2015
Photographs: Courtesy of Primus Architects

The new library and culture centre is an extension and refurbishment of the former Fritz Hansen furniture factory. The project marks the beginning of a larger transformation of the city centre of Allerød, north of Copenhagen, which includes new urban spaces and an extension of the Mungo Park Theatre.

The extension is divided into two spaces. One serves as a cultural centre and can be subdivided into a theatre/event area and a community space. The other is directly connected to the existing library and houses the childrens library. The shedded roof both provides amplient space and natural light and supportingthe acoustic climate of the interior space. At the same time it aknowledges the industrial history of the site. The angle of the roof gives optimal use of the integrated solar panels.

Section 1

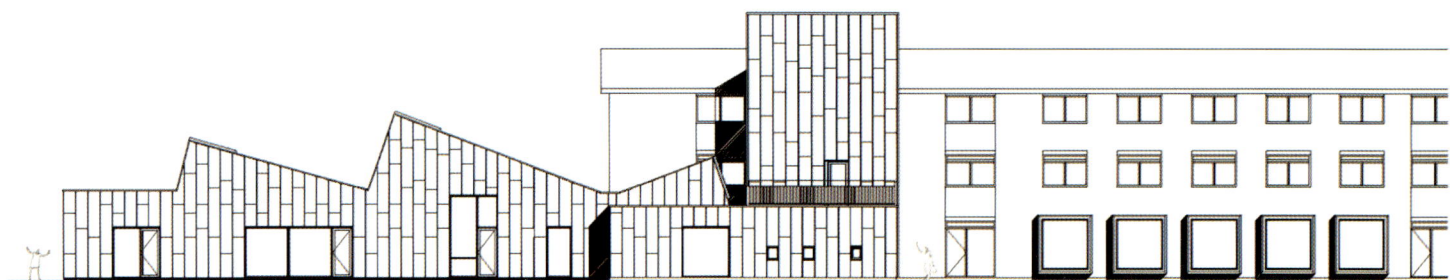

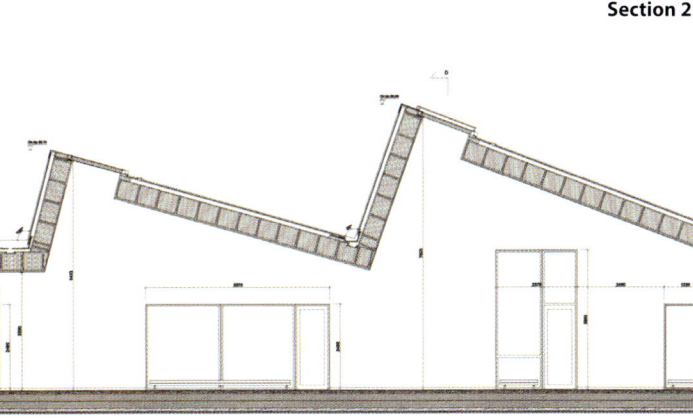

Section 2

278

Sketch

Floor Plan

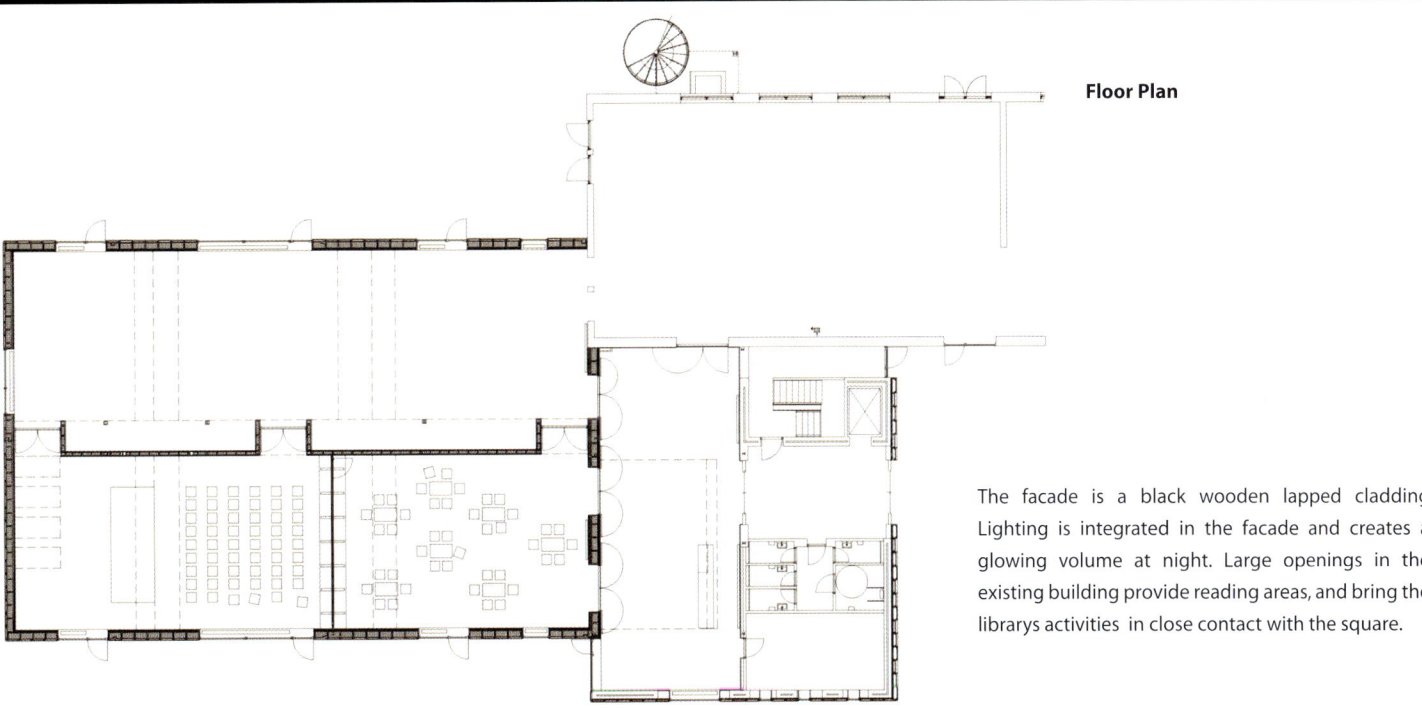

The facade is a black wooden lapped cladding Lighting is integrated in the facade and creates a glowing volume at night. Large openings in the existing building provide reading areas, and bring the librarys activities in close contact with the square.

Len Lye Centre

Architects: Patterson Associates
Location: New Plymouth, New Zealand
Project Year: 2015
Photographs: Patrick Reynolds

It was Lye himself who said in 1964 that "great architecture goes fifty-fifty with great art," a maxim that has informed the approach and form of the Patterson Associates-designed Antipodean Temple that houses his work.

Lye was fascinated with temples and in conceiving the overall design it seemed aesthetically and historically appropriate to draw inspiration from the "megarons," or great halls, of the classical world, as well as Polynesian forms and ideas. These also influenced Lye and he is, after all, the client. To do this in a new way, we developed our thinking in a holistic or adaptive way, using what we call "systems methodology." This means that rather than using proportion or aesthetics, we use patterns in the ecology of the project's environments to drive the design elements.

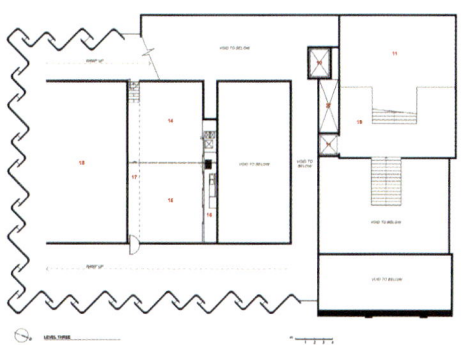

For example, the shimmering, iridescent colonnade façade, manufactured locally using stainless steel - Taranaki's 'local stone' - links both Lye's innovations in kinetics and light as well as the region's industrial innovation. By doing this we celebrate the fortunate gift of his works to Taranaki. The colonnade creates a theatre curtain, but with three asymmetric ramped sides, leading to a type of vestibule, known as "pronaos" in Ancient Greece. This is formed by the gallery holding the large Lye works. Viewed from above, the colonnade's top edges create a koru form, displaying the Museum's Polynesian influences as the meeting house, or wharenui, for Len Lye. The procession of the colonnade morphs into a portico, announcing the main gallery as a type of megaron but also functioning as a wharenui; the deities and ancestors referenced and represented by Lye's inspirational work.

Traditionally, the most sacred and private part of a temple, the "adyton," is located at the point furthest from the entrance. Here is housed the Len Lye archive, while the 'treasury,' known as the "opisthodomos," looks back to the people entering below. The project respectfully links into the smaller existing Govett Brewster Art Gallery, which itself has been retrofitted from the city's decommissioned heritage cinema. The combined facility is undivided, with a circular loop allowing visitors to appreciate the changing museum and gallery displays within one flexible and shared structure.

Lebourgneuf Community Center

Architects: CCM2 architectes
Location: Québec, Quebec City, QC G1K, Canada
Area: 2330.0 sqm
Project Year: 2014
Photographs: Stéphane Groleau, Dave Tremblay

Structure / Civil: EMS
Mechanical / Electrical: WSP Group
Contractor: Construction Béland-Lapointe inc.
Budget: 5,6 M$ | CAN

Located in Quebec City, the Lebourgneuf Community Center is a central element for the community because of the activities offered and its proximity to Les Prés-Verts Elementary School. The Center is directly connected to the school and is a major partner for recreational, educative and sociocultural activities. The extension project consisted in adding a double gymnasium, two locker rooms with sanitary facilities, a divisible workout gym and a classroom.
The extension has two steel structure levels and was designed according to a future extension of Les Prés-Verts Elementary School and a new local library.

The concept has been developed considering many constraints of the site such as a wooded area, liable to flooding areas and future constructions (school and local library). The extension is a pilot study for universal accessibility.
Every public space, circulating area, locker room and class room is designed in order to allow people with reduced mobility, impaired vision, elderly person or young children to be independent in the building.

IDENTIFICATION MURALE
CONTRASTE ÉLEVÉ ET TYPOGRAPHIE FACILITANT LA LECTURE

PORTES PUBLIQUES
PORTES DE COULEUR CONTRASTANTE POUR FACILITER LE REPÉRAGE

SIGNALISATION
SIGNALISATION CONTRASTANTE ET INSTALLATION PERPENDICULAIRE AU MUR POUR FACILITER LE REPÉRAGE

MAIN-COURANTE
MAIN-COURANTE RONDE FACILITANT LA PRÉHENSION

GESTION DES NIVEAUX
CRÉATION D'UN PLAN INCLINÉ AVEC PENTE INFÉRIEURE À 1:20 + PALIER INTERMÉDIAIRE (SURFACE ANTIDÉRAPANTE)

BANDE TACTILE
BANDE AU SOL FACILITANT LE REPÉRAGE POUR LES PERSONNES À VISIBILITÉ RÉDUITE (CANNE BLANCHE)

MOBILIER D'ACCUEIL
HAUTEUR ADAPTÉE + ALCÔVE CONTRASTANTE POUR FAUTEUIL ROULANT

ESCALIER
NEZ DE MARCHES ET CONTREMARCHES CONTRASTANTES

GARDE-CORPS ET MAIN-COURANTE CONTINUE
POUR FACILITER LE CHEMINEMENT DES PERSONNES À MOBILITÉ RÉDUITE (VISUELLE ET MOTRICE)

LIGNAGE AU SOL
LIGNE DE CONTRASTE AU SOL ANNONÇANT LA PROXIMITÉ DE L'ESCALIER

IDENTIFICATION AU SOL
GYMNASE (NOIR)
VESTIAIRES (JAUNE)

Learning Hub

Architects: Heatherwick Studio
Location: Nanyang Technological University, Singapore
Area: 14000.0 sqm
Project Year: 2015
Photographs: Hufton and Crow

Environmental Performance: Green Mark Platinum
Design Consultant: Heatherwick Studio; Project Lead - Ole Smith
Lead Architect: CPG Consultants; Project Lead - Vivien Leong
Civil & Structural Engineers: TYLin International

Founder and Principal Thomas Heatherwick, Heatherwick Studio says, "Heatherwick Studio's first major new building in Asia has offered us an extraordinary opportunity to rethink the traditional university building. In the information age the most important commodity on a campus is social space to meet and bump into and learn from each other. The Learning Hub is a collection of handmade concrete towers surrounding a central space that brings everyone together, interspersed with nooks, balconies and gardens for informal collaborative learning. We are honoured to have had the chance to work with this forward-thinking and ambitious academic institution to realise such an unusual project."

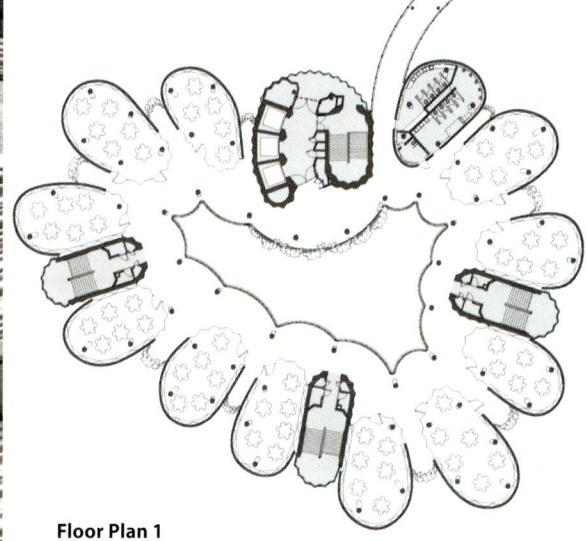

Floor Plan 1

Project lead Vivien Leong of CPG Consultants, the Lead Architect and Sustainability Consultants for the Learning Hub, says, "The most exciting aspect of this project is to see such an inspired design develop into a uniquely contextual and functional building through a highly collaborative process. Managing this project was no mean feat as we had… …to ensure that our work complied with Singapore's rigorous building regulations and that it achieved the highest standards of sustainability, while working hard to retain the integrity of the original design and vision of NTU . The opportunity to challenge convention by introducing several first of its kind environmentally friendly features and innovative solutions that embody the spirit of modern day learning has been a truly rewarding experience for us." The combination of local building codes and high environmental aspirations meant that a concrete construction was necessary. The primary design challenge was how to make this humble material feel beautiful.

As a result, the concrete stair and elevator cores have been embedded with 700 specially commissioned drawings, three-dimensionally cast into the concrete, referencing everything from science to art and literature. Overlapping images, specially commissioned from illustrator Sara Fanelli, are deliberately ambiguous thought triggers, designed to leave space for the imagination. The sixty one angled concrete columns have a distinctive undulating texture developed specially for the project. The curved facade panels are cast with a unique horizontal pattern, made with ten cost-efficient adjustable silicone moulds, to create a complex three-dimensional texture. The result of the building's various raw treatments of concrete is that the whole project appears to have been handmade from wet clay.

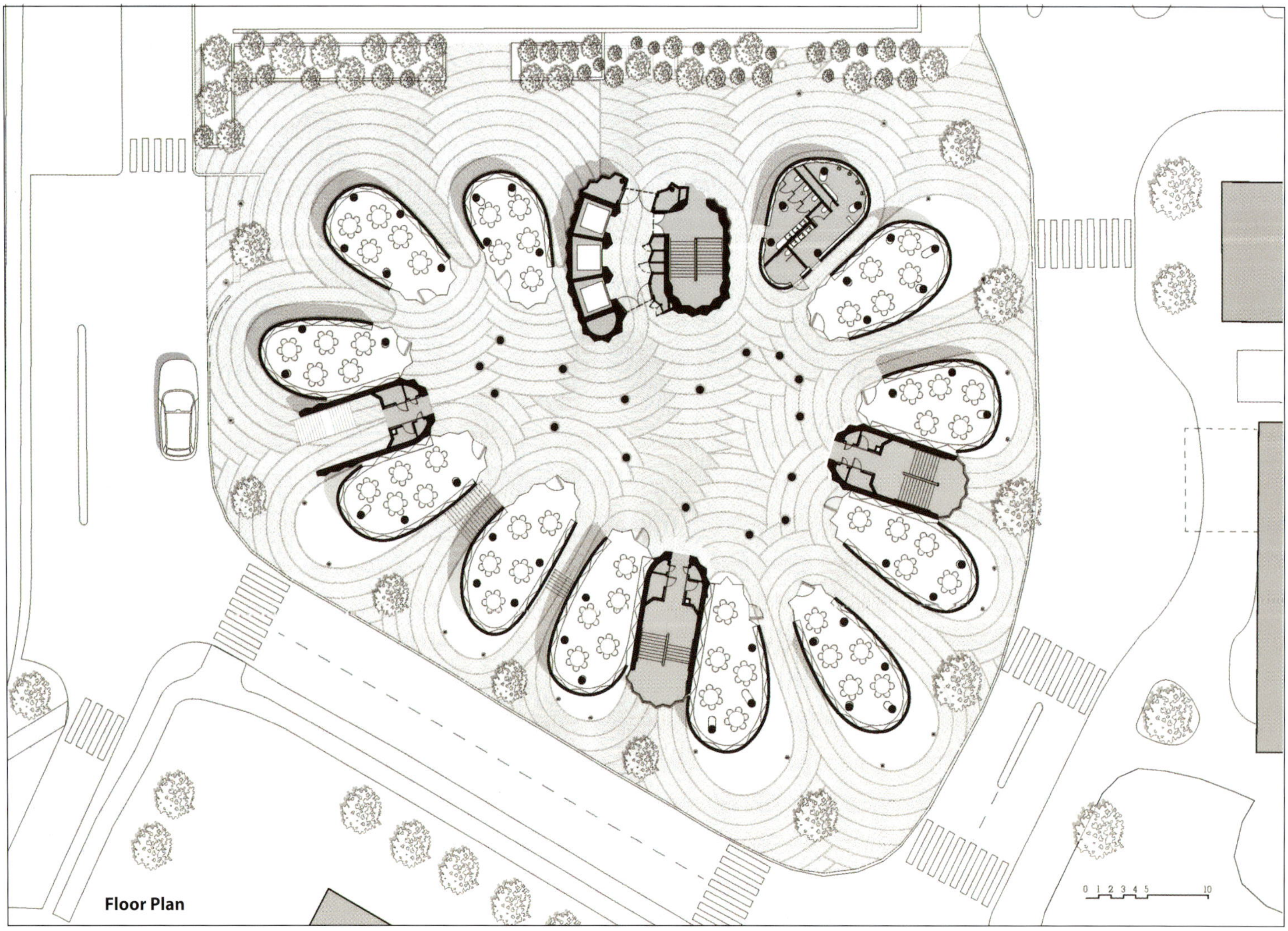

Floor Plan

Le Cristal Cinema and Michel Crespin Square

Architects: Linéaire A
Location: Aurillac, France
Area: 6850.0 sqm
Project Year: 2015
Photographs: Hervé Abbadie

Architect in Charge: Castelbajac, Deby, Makarem
Structure: Sibeo Ingenierie
Acoustics: Peutz
Scenographer: Linéaire A - Castelbajac, Deby, Makarem

Completing the fourth side of a square comprised of three historic barracks buildings, the layout of our cinema defines a spacious and legible square, offering a flexible new public space for hosting cultural events such as the Aurillac's international street theatre festival. Compact and efficient in its function, the Cristal Cinema wraps around the full height entrance hall, which peels away to enhance the view between Carmes Street and the clock building. The project's sculptural expression implies the motion inherent to cinema and was envisaged as a rock crystal grown out of the square, its facets continuously playing with light throughout the day and night.

294

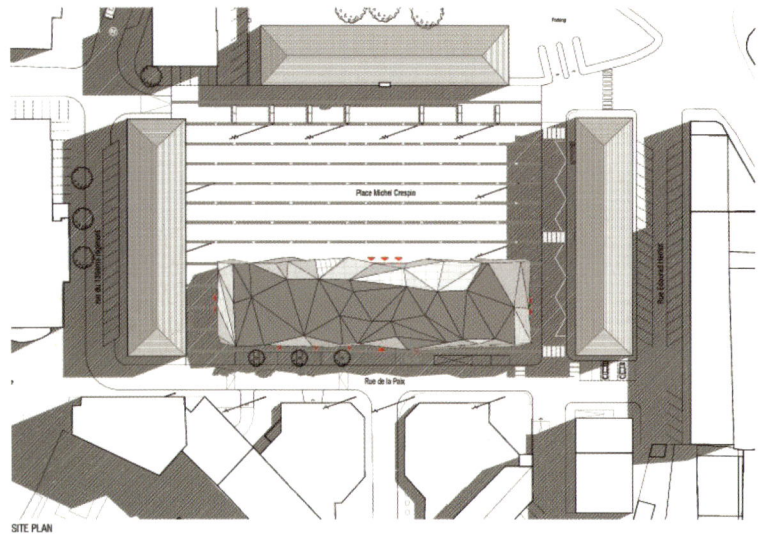
SITE PLAN

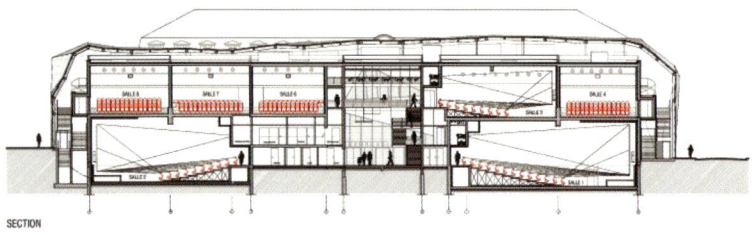
SECTION

SECTION

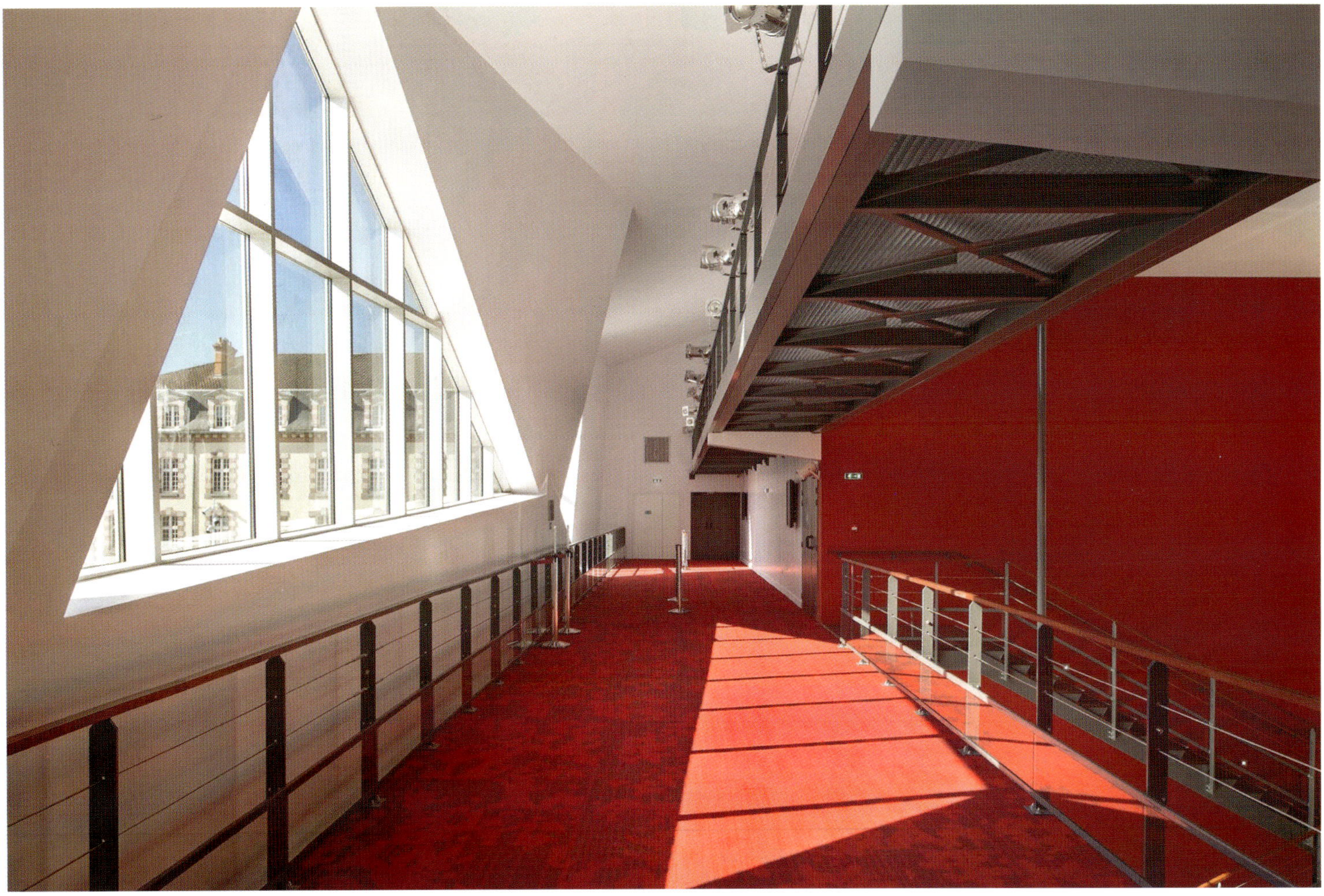

296

École Nationale Supérieure Maritime in Le Havre

Architects: AIA Associés
Location: Le Havre, France
Area: 4500.0 sqm
Project Year: 2015
Photographs: Luc Boegly

This urban development is anchored in the relationship between the city and its docks; its morphology, its character and its texture all bring to mind ocean-faring vessels. Its position, parallel to the quay, places it in a direct relationship with the basin and the port. Over a 100m stretch along the quay, the building grazes the waterfront on one side while stretching towards the city on the other. From the entrance to Le Havre, its prow cuts a lean figure, rising towards the city. It stands out among the initial views of the port as visitors enter the city along the Vauban basin. Set between earth, sky and sea, the school seen from afar suggests a ship on the high seas, but also the great selachians of the deeps.

ENSM

NORD
QUAI
0 5 10 15 20

COUPE LONGITUDINALE N-S

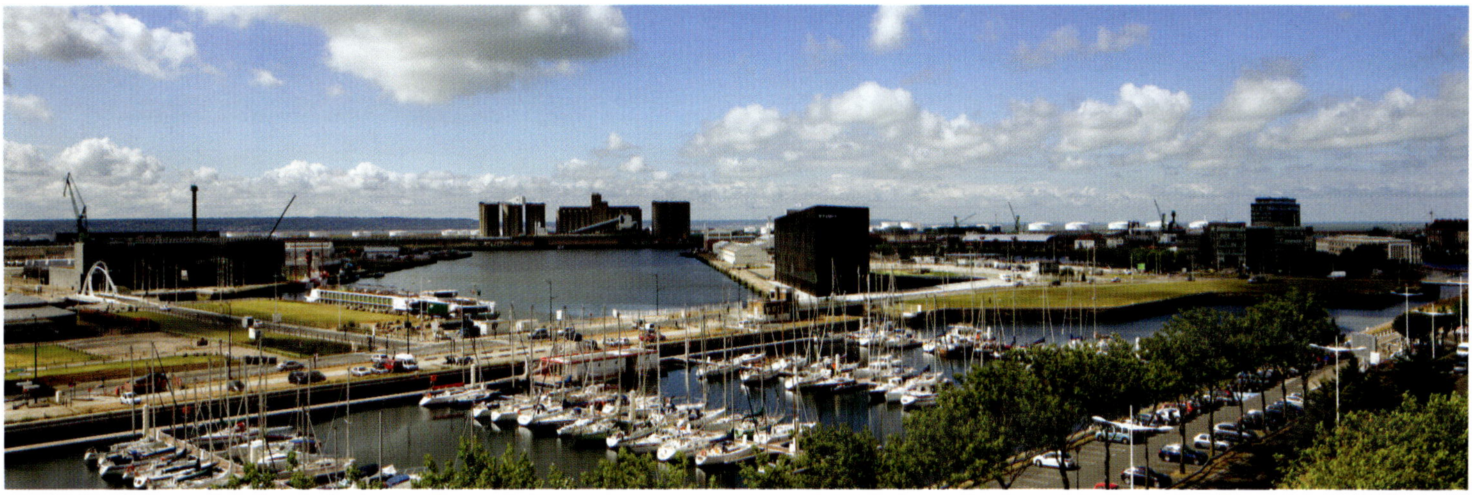

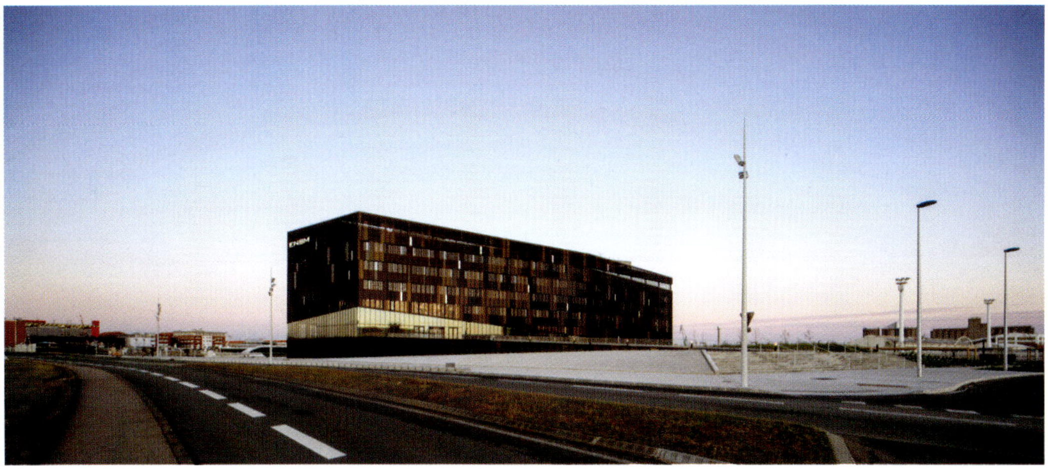

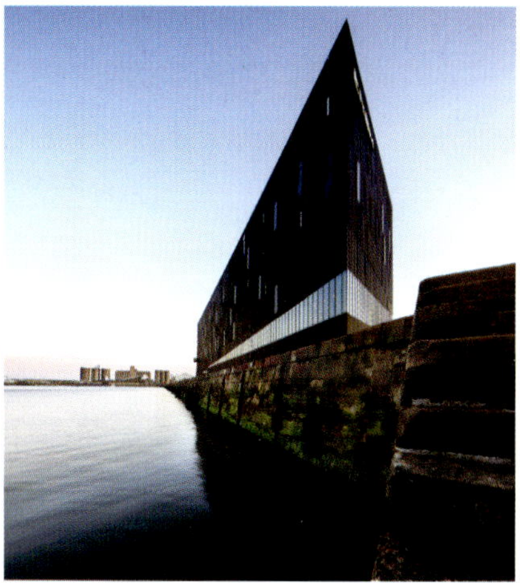

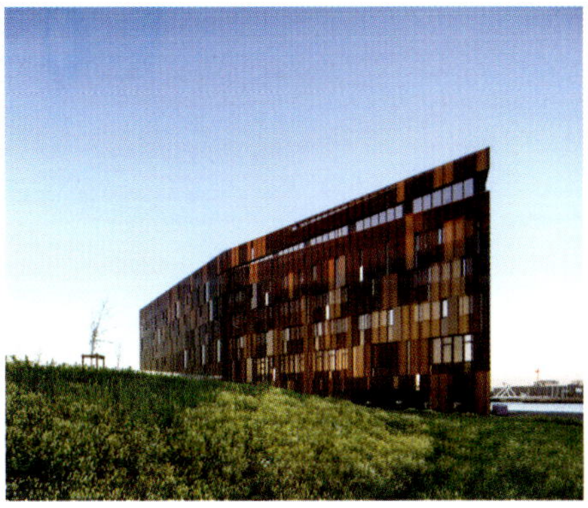

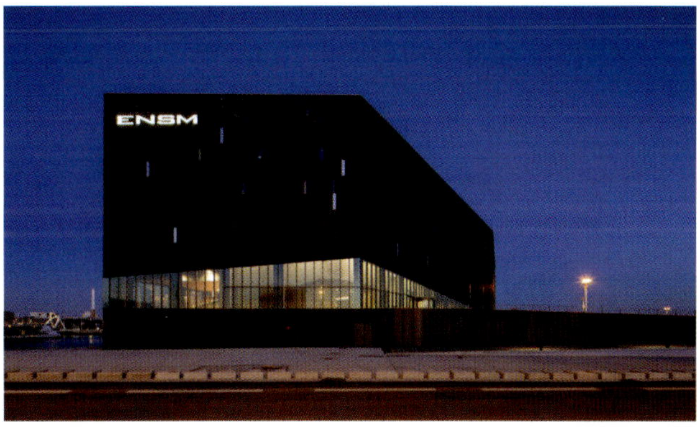

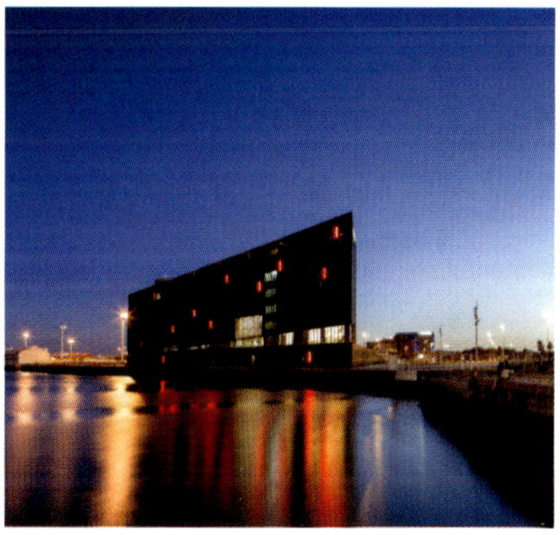

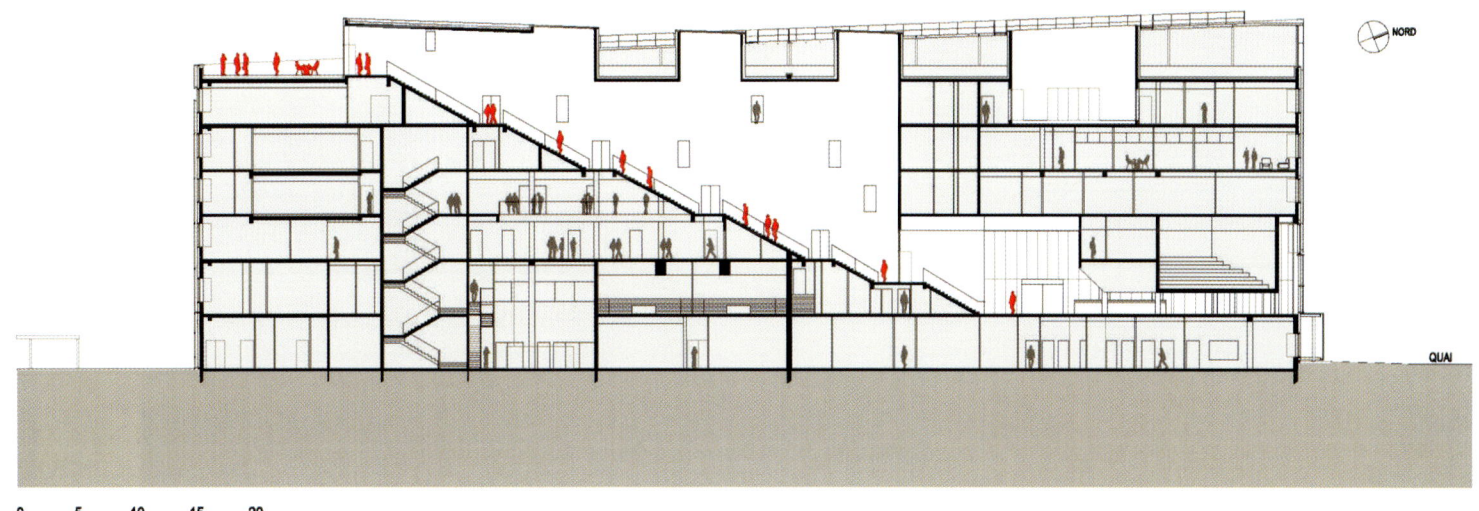

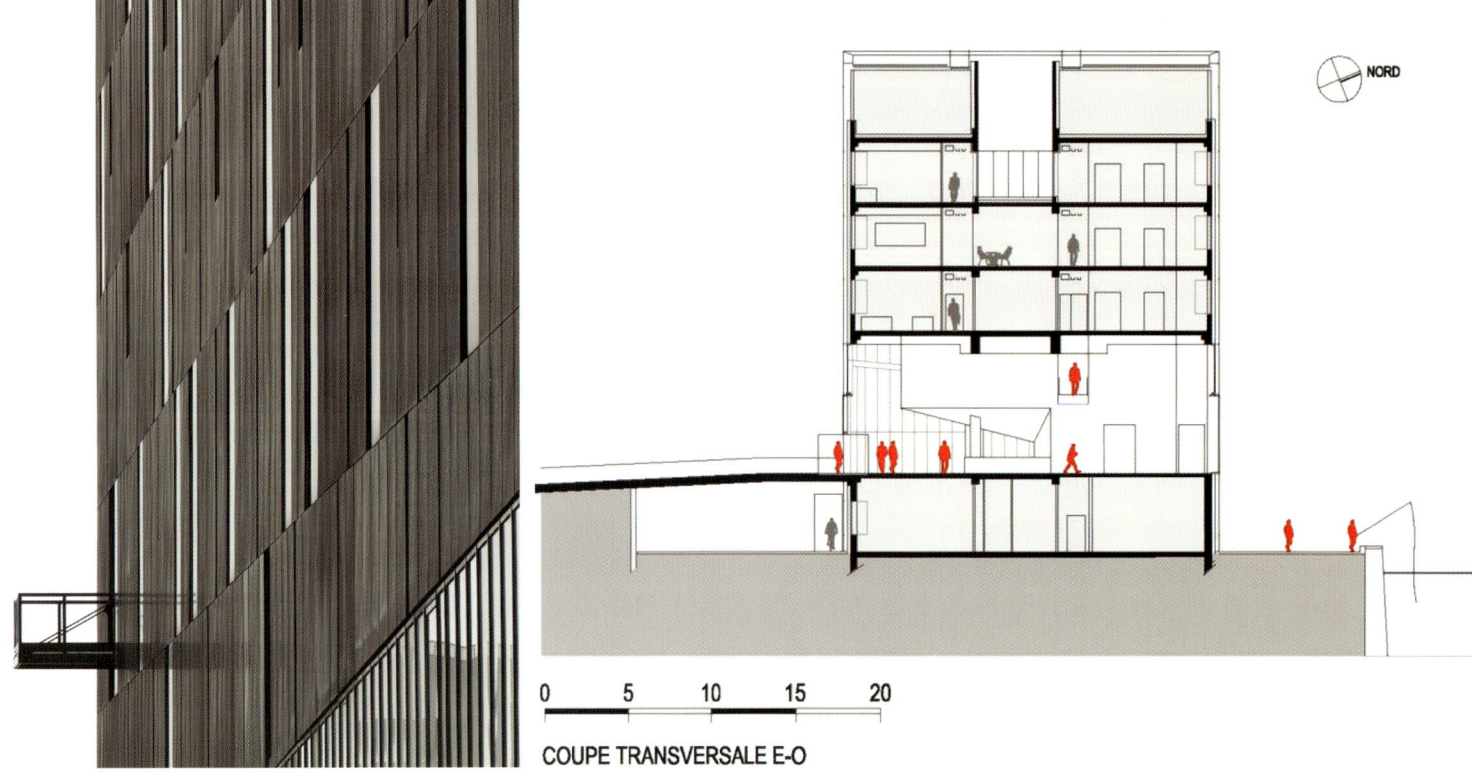

COUPE TRANSVERSALE E-O

CZECH REPUBLIC PAVILION EXPO 2015

Architects: Chybik+Kristof Architects & Urban Designers
Location: Padiglione Svizzera - Expo 2015, Via Belgioioso Cristina, 70, 20157 Milano, Italy
Area: 3200.0 sqm
Project Year: 2015
Photographs: Lukas Pelech

Through the end of October, the Czech Republic, along with over 150 other counties of the world, is participating in the Expo 2015 in the northern Italian metropolis of Milan. Expo, far beyond being a venue to showcase progress and the latest discoveries in science and technology, also aims to contribute to solving current global problems and to strengthen international cooperation. The motto of the Milan Expo is: "Feeding the Planet, Energy for Life", referring to the need for balanced securing of quality sources of food and potable water for everyone on the planet. The Czech Republic chose water as the central theme of its exposition, since in the European Water Charter of 1968 it is stated that "water is indispensable to all forms of life" and its preservation "is the joint responsibility of states and all users".

Section

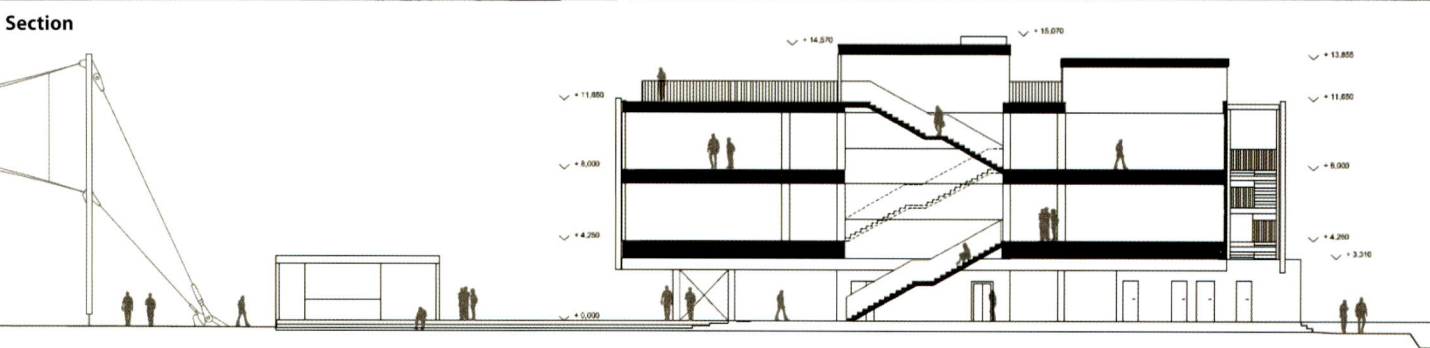

Keynsham Civic Centre

Architects: AHR
Location: 1 Market Walk, Keynsham, BS31 1FS
Area: 78000.0 ft2
Project Year: 2015
Photographs: Courtesy of AHR
Structural Engineer: Hydrock
Environmental Engineer: Max Fordham
Landscape Designer: Novell Tullett / AHR
Acoustics Consultant: Max Fordham / Mach Acoustics
Main Contractor: Willmott Dixon

The new development has replaced 1960s buildings to provide 68,000sqft council offices, a library and one-stop shop, 20,000sqft retail, two new pedestrian streets, a market square, car parking and highways improvements.

AHR responded to the challenge of the constrained and sloping town centre site creating two new pedestrian streets and a cluster of inter-locking buildings. This allowed over 50% of the site to be given over to new public realm and better integrated the development into the existing urban grain. Careful orientation of the buildings and an innovative acoustic louvre window system allowed full natural ventilation to be used throughout, despite the location within a busy town centre.

At the heart of the brief was Bath & North East Somerset Council's objective of creating a highly efficient, robust and flexible building that would minimise energy consumption and maintenance while providing first class civic facilities and a high quality workplace embodying a 'one council' culture.

Detail

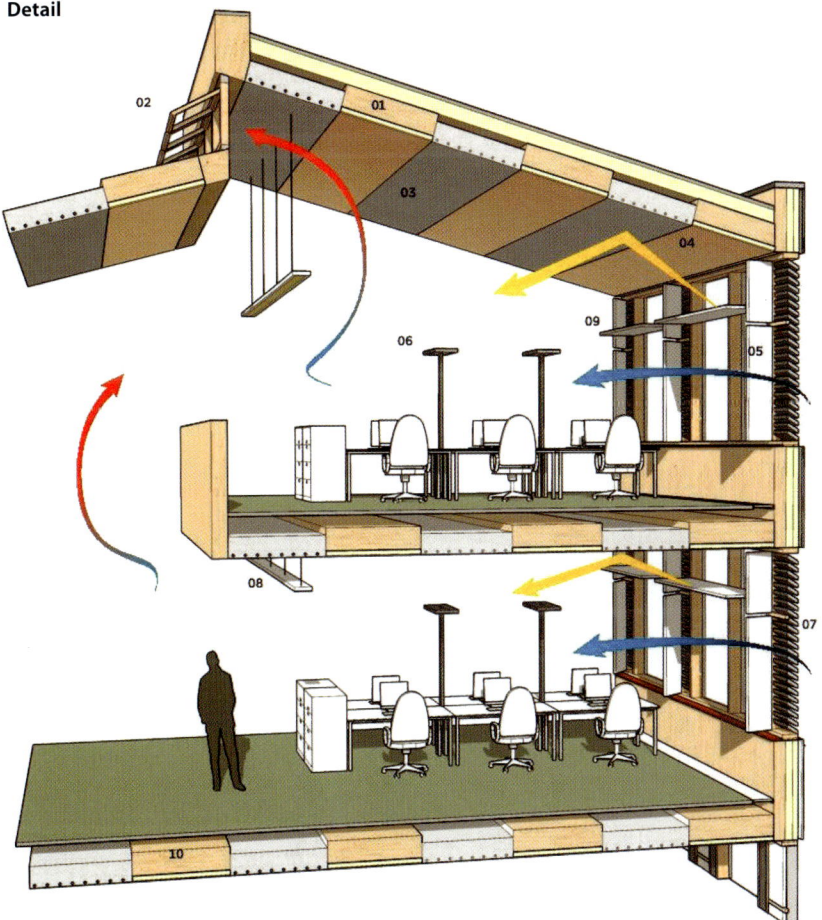

AHR has built on extensive in-house research and worked closely with Environmental Engineer Max Fordham to develop a pioneering environmental strategy for the project: it is the first in the UK to implement BSRIA Soft Landings to target an exemplary DEC A rating, which it is expected to achieve in 2017 once it has been in use for two years. This rating will confirm Keynsham Civic Centre as one of the lowest energy consuming public buildings in the UK. The offices have already achieved an outstanding EPC A rating and the project is almost carbon neutral. Every aspect of the design of the building was considered to provide optimal environmental performance, starting with the building's orientation and form which reduced the requirements for mechanical ventilation and lighting. An innovative timber (CLT), steel and concrete structure not only created bright and welcoming interiors but also reduced heating requirements and construction time.

Keynsham Civic Centre has provided Bath & North East Somerset Council with a landmark building which positions it at the heart of the community it serves and which embodies its commitment to that community and to its staff to be an exemplary Council, investing wisely, making a positive impact in every aspect of its work and looking to the future.

The Broad Museum

Architects: Diller Scofidio + Renfro
Location: 221 South Grand Avenue, Los Angeles, CA 90012, USA
Area: 120000.0 ft2   **Total Cost:** $140 million
Project Year: 2015
Photographs: Benny Chan, Iwan Baan, Jeff Duran - Warren Air

The Broad is a new contemporary art museum built by philanthropists Eli and Edythe Broad on Grand Avenue in downtown Los Angeles. The museum, which was designed by Diller Scofidio + Renfro, will soon be open. The museum will be home to the nearly 2,000 works of art in The Broad Art Foundation and the Broads' personal collections, which are among the most prominent holdings of postwar and contemporary art worldwide. With its innovative "veil-and- vault" concept, the 120,000-square-foot, $140-million building will feature two floors of gallery space to showcase The Broad's comprehensive collections and will be the headquarters of The Broad Art Foundation's worldwide lending library. The Broad is also building a 24,000-square-foot public plaza adjacent to the museum to add another parcel of critical green space to Grand Avenue.

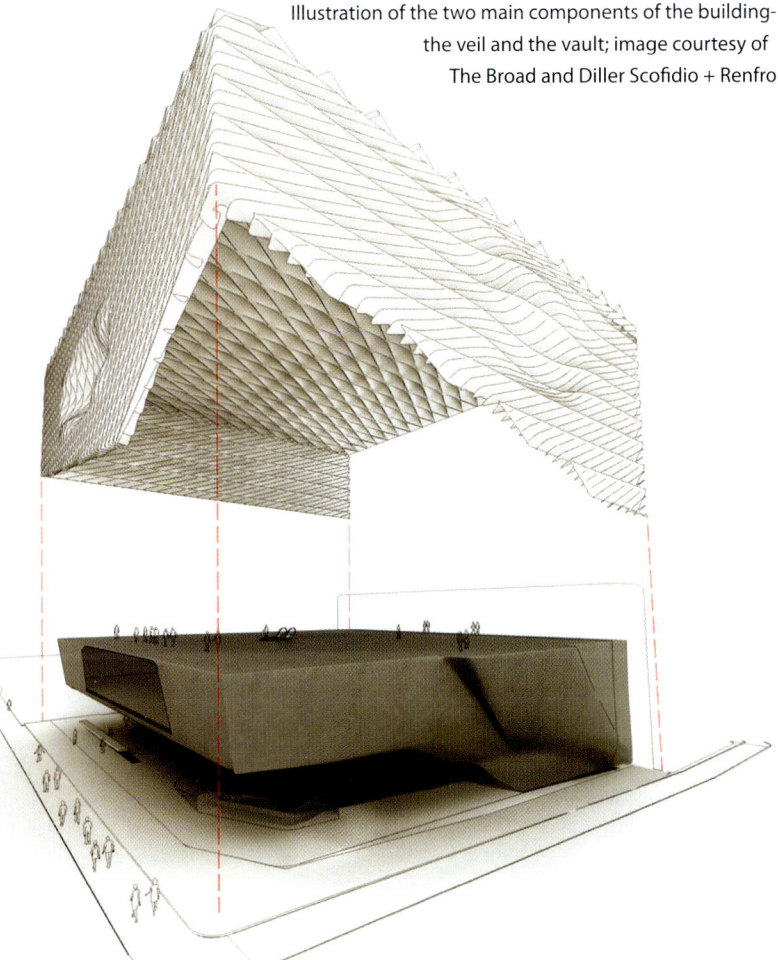

Illustration of the two main components of the building- the veil and the vault; image courtesy of The Broad and Diller Scofidio + Renfro

SF Tower, USA

Architects: HNTB Architecture
Location: San Francisco, CA, USA
Design Team: Joe Grogan, Paul Kim, Alejandro Ogata, Robert Steel
Project Year: 2015
Photographs: Courtesy of HNTB Architecture

What makes an iconic design? This was the first question we asked ourselves at the beginning of the design process. This was going to be, after all, a highly visible structure at one of the most iconic airports in the United States at one of them most design-oriented cities in the world.

The goals of both the Airport and the Architect were to reflect the SFO brand with an elegant form and welcoming gesture to the public. This is directly manifested in the sweeping torch form and the impetus for opening of the tower shell to expose the 'core' with a back lit glass facade which greets passengers arriving along the upper level roadway and train. The intent was to provide a sleek and elegant façade with a subtle but dynamic skin with spiraling aluminum panel joints around the conical surface.

At the same time we saw an opportunity on the location of the tower. Unlike towers at most airports, SFO's is immediately adjacent to drivers and pedestrians. We wanted people to experience the tower up close as well as from afar. The tower comes down on to a glass box at a public connector between terminals 1 and 2. This moves makes the tower appear taller from the roadway while providing passengers dramatic views from the base of the tower.

Though effortless in appearance, the elegant form shrouds a feat in engineering and functional design. The aluminum skin conceals 75,000 pounds of steel weights to resist wind and earthquakes as well as the large amounts of high tech equipment needed to run a modern airport terminal.

NEW BUILDING ELEMENTS

1. **Airport Traffic Control Tower (ATCT)** to replace the existing ATCT
2. **Integrated Facility** base building that includes FAA and Airport functions
3. **Non-secure Corridor** for passengers circulating between Terminals 1 and 2
4. **Secure Connector** for passengers circulating between Terminal 1 and 2 boarding areas

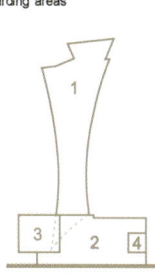
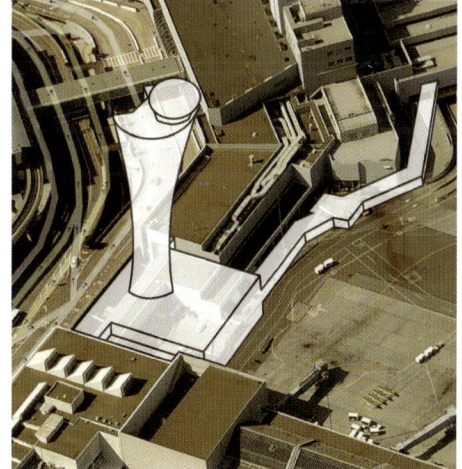

310

2016 공간구성과 트랜드 I

공간의 조형미와 최신 건축자재의 트랜드

2016년 4월 1일 발행

- 출　판　　**아르스**
- 발행인　　**박 제 하**
- 발행처　　**아르스**

　　　　　경기도 고양시 일산 동구 장항동 552-10
　　　　　대표전화: 031)907-1777　Fax: 031)907-1777

- ISBN: 979--11-86338-35-3 (SET)
- ISBN: 979--11-86338-38-4

Printed in Korea

정가: 95,000원